THE STARS BURN ON

Also by Denise Robertson

The land of lost content (1985)
A year of winter (1986)
Blue remembered hills (1987)
The land of lost content: the Belgate trilogy (1988)
The second wife (1988)
None to make you cry (1989)
Remember the moment (1990)

THE STARS BURN ON

Denise Robertson

Constable · London

First published in Great Britain 1991
by Constable and Company Limited
3 The Lanchesters, 162 Fulham Palace Road
London W6 9ER
Copyright © Denise Robertson 1991
The right of Denise Robertson to be
identified as the author of this Work
has been asserted by her in accordance
with the Copyright, Designs and Patents Act 1988
ISBN 0 09 469560 1
Set in Linotron 11pt Palatino by
CentraCet, Cambridge
Printed in Great Britain by
St Edmundsbury Press Limited
Bury St Edmunds, Suffolk

A CIP catalogue record for this book
is available from the British Library

Prologue

6.50 a.m. New Year's Day
1980

They raced one another to the summit, stumbling over the tussocky grass in the darkness, laughing through teeth that chattered with cold, buoyed up by booze and youth and Auld Lang Syne.

And then they were on top of the hill, the wind plucking at them as they looked down on the brilliant nightface of Sunderland to the east, and turned to see the dark acres of Durham County to the west.

'There it is then – our roots. You wanted to see it. I hope you're satisfied.'

'I will be when I get this cork out.' There was a plop and then red wine was passing from hand to hand.

'Here's to the new decade,' someone said. 'May the next ten years be good to us all.'

'I'm freezing,' a girl complained. 'Can't we get out of the wind? Whose idea was this, anyway?'

'Cuddle in.' One of the boys was opening his long serge coat to enfold her. 'Better?'

'Yeah. Ta.'

There were eight of them there, all friends. Now they moved down from the summit, into the lee of the wind, drawing closer for warmth, wiping the neck of the bottle in turn and swigging the wine.

'Where will we be next year? Not all together, that's for sure.'

A belch and a chuckle. 'Cheer up! This time next year most of us will be earning – cling to that.'

'There are eight of us,' said the voice of authority. 'Statistically . . .' There was a groan but the voice was undeterred. 'Statistically, one of us will be dead, in a road accident probably, three will be married . . .'

5

'Me,' someone said and squeezed the girl inside his coat until she gasped, her breath a small cloud on the night air.

'. . . and two of us will still be looking for jobs,' the voice went on. 'If you don't count the one who's already earning, that means only one of us has a chance of being even moderately rich. And that bugger inherited it.'

'God, you talk like a politician already,' a girl said. 'Who's got the bottle?'

'It's dead, I'm afraid.' This voice was less northern than the rest, and apologetic. 'I've got some sangria and some gin here, if that's any good.'

They passed the bottles round, all drinking sparingly, anxious to share, one girl abstaining as she had done all night because she was driving her father's car and had given her word.

It was time for another toast. 'To freedom,' a girl said. 'To the year we all escape!' Suddenly there was shared apprehension at the thought of possible failure.

'Cheer up,' the bringer of gin and sangria said, sensing a hush. 'You'll all do brilliantly and conquer the world and I shall bask in your glory, having none of my own.'

'Pass the bottle,' said another. He raised it until it could be seen against the night sky. 'This is a libation to placate the gods of the hill. May they pour blessings on us all . . . even the silly sod who had the misfortune to be born a member of the ruling classes. And may we all reach those positions of power and affluence for which our superior intellects have equipped us.' There was a crackling noise as the sangria fell on to the frosted grass, and somewhere in the darkness a bird began cheeping.

'Let's meet here in ten years, then we can see what's happened to us all.'

'You won't get me up here again in mid-winter,' a girl said firmly.

'Make it midsummer then: the twenty-first of June 1989, at 8 p.m. On this very spot.'

No one spoke.

'OK?' the proposer demanded.

There was a murmur of agreement.

'Right, now shut up. There's a mystic light in the east, or

6

there bloody well should be, and I want to commune with nature.'

'He thinks he's a Druid,' one of the girls said, but she too fell silent as the sky slowly lightened to reveal tower-blocks and spires, and the birds began to sing as though their lives depended on their song.

BOOK ONE
1980

1

New Year's Day

They stayed on the hill while the lights of the town grew pale by comparison and the birds became frenzied. The houses were still sleeping but here and there a lighted Christmas tree glowed in an uncurtained window. 'The houses are stealing the hillside,' Jenny thought, remembering childhood and waist-high grass where now there were driveways and patios.

'No rosy-fingered Aurora,' Alan said ruefully. It was true: there was no obvious source of light, but everything was taking on an incandescence. They were all shivering, the excesses of the night before beginning to tell, but they went on standing there, as though in awe, until the birds accepted day as a fact and ceased their song.

'Come on,' Keir said. 'I'm frozen.' They moved back over the summit and down towards the cars, stamping their feet on the frozen grass to restore circulation, blowing into their hands, each boy turning occasionally to help the nearest girl over a particularly steep patch.

'What must it be like in Afghanistan?' Jed said, suddenly serious. It was a week since the Russians had seized Kabul airport and their reinforcements had rolled across Afghanistan's northern border.

'They'll fight in the hills,' Alan said. 'Remember, we never managed to beat the Afghans.'

'We used kid gloves,' Euan said. '*Noblesse oblige* and all that. I doubt whether the Red Army has the same scruples.'

'Who taught you history?' Keir asked drily. 'Rudyard Kipling, from the sound of it. We were bastards when we were a colonial power. Don't believe anything else.'

'Not on New Year's Day, please, gentlemen.' Kath was

11

turning back to glare. 'We don't want to be patronized today, Keir, thanks a bunch. We're not up to it.'

Somewhere in the distance there was the banging of car doors and shouted farewells. 'Someone else has had enough,' Euan said, and then, as they neared the allotments, 'I'm famished. Let's forage for food.'

'What food?' Barbara said.

'I don't know. Turnips? Maybe there's a hot-house there with grapes?'

'Silly bugger,' Kath answered amiably. 'They're allotments.'

'I was only joking,' Euan said.

'Can we get a move on?' Elaine's face was pinched, her eyes blue-ringed beneath mascara-clogged lashes.

'She looks awful,' Jenny thought, 'and at the same time beautiful.'

'Who's with me, then?' Euan asked, jingling his car keys, but Jed had turned back to look at the hill.

'Christ,' he said, 'did we climb that?'

'It's only a little hill.' Barbara was unlocking the door of her father's Escort. 'Come on, let's get organized.'

Jed was drawing Elaine to him, folding her into his coat again, but his eyes were on Barbara, the old animosity there, the chin with its new day's shadow beginning to jut. 'You know your trouble, Babs?'

'Don't call me that! My name's Barbara.'

'You know your trouble?' Jed persisted. 'You are stinking sober. Utterly, horribly un-inebriated. Your judgement's gone, your reflexes are shot to pieces, you are seeped in citrus juice and your liver . . . God, your liver . . .'

'Shut up, pillock,' Alan said, opening the car door wider. 'If she's sober it's because she's been driving all night. Be grateful.'

'So has Euan,' Jed said, 'and he's managed to get pissed.'

'Yes,' Alan said. 'Let's hope he gets home without being nicked. Now, who's for Barbara's car?'

Jenny felt a glow of approval. If her instincts were right, Alan didn't like Barbara but he was being fair. He would make a good barrister. Her eyes flicked to Keir and then quickly away, afraid to meet his gaze in case he saw her love for him writ large on her face.

They separated then – Alan, Jenny and Elaine into Barbara's Ford because, like her, they lived in town; Jed, Keir and Kath into Euan's Rover because he was driving them out to Belgate, their mining village.

The two cars bumped along the track to the road and then down towards the town, Euan leading the way.

'Look at that!' Barbara said, suddenly scandalized. The car ahead had a sun-roof and through it Jed had emerged, naked to the waist except for his college scarf.

'What's he doing?' Alan chuckled from the front seat. Jed was bowing left and right to the still-sleeping houses, moving his right arm in a royal gesture.

'Making an exhibition of himself, as usual.' Barbara changed gear with a venomous jerk.

'He'll get the car stopped if he's not careful.' Alan was making signs through the windscreen that Jed should desist. There was movement in the other car, too, and then Jed was pulled struggling from sight.

'Thank goodness they turn off here,' Barbara said as they reached the roundabout. The Rover went right, towards the county and the Escort drove on towards the town centre.

'He's canny, Jed, isn't he?' Elaine said. Close to her in the back seat Jenny could see how thin Elaine was, how pallid was the perfect skin, but her mouth had curved with pleasure as it always did at the mention of Jed.

'Yes,' she said. 'He's a nut-case but he means well.'

They dropped Alan at the end of his road. 'Usual place tonight?' Barbara asked, as he made to shut the door. Alan rolled his eyes.

'I'm not sure I'm up to it. I'll see.'

'Come if you can,' Barbara coaxed. 'You'll be out, won't you Jenny?'

She was pointedly ignoring Elaine and Jenny prickled with embarrassment. 'I don't know,' she answered.

'Let's all sleep on it.' Alan was already closing the door. 'Thanks for the lift.'

'You don't have far to go, do you?' Barbara said to Elaine firmly, knowing her flat was quite some distance away but stopping at the next junction anyway.

'Ta,' Elaine called as she tottered off on her spindly heels.

13

'I'm working tonight. Pop in if you can.' Suddenly she reached down and slipped off the flimsy court shoes, turning to grin and wave at the watchers before she sped barefoot over the icy pavement.

'She's so common,' Barbara said as they turned back towards the hill.

'She's so pretty,' Jenny declared fervently. 'Beautiful, even.'

'I don't know why she hangs on to us, though.'

'Because the boys are gone on her,' Jenny said ruefully.

'Who is?'

'Jed for one. *And* Euan. *And* Keir.'

'And we know why!' Barbara said.

Jenny closed her eyes in vexation. 'No we don't, Ba, so we shouldn't say it.'

'Well, anyway, she has her job in the café so she doesn't have to be a barmaid. She must like the life. It wouldn't do for me.'

'She does it for the money,' Jenny said. 'She doesn't have doting parents like you and me. She needs two jobs.'

There were cars on the road now and the odd pedestrian walking his dog. It was a relief to Jenny to see her own gate and escape into the cold, clean air. 'I don't know if I'll come out tonight, but you can ring,' she said and Barbara made a face.

'Come if you can, Jen.'

Jenny was beginning to doubt her capacity to carouse again. All she could think of was bed, sinking into smooth sheets, curling up for warmth and letting go. Still, Barbara had been her friend since Infants' School. 'I'll try,' she promised.

She went up the neat path, between savagely pruned rose-bushes, past the tiny ornamental pool with its shell ornaments. The garden was her father's pride and joy but it was too neat, especially in winter when black earth showed stark between the planted rows. He kept the garden as her mother kept the house, immaculate.

Inside the house she moved slowly, terrified lest she wake the dog and his barking should alert her mother. She took her shoes off in the hall and avoided the creaking stair, but all in vain.

14

'Is that you, Jennifer?'

She leaned her head against her mother's bedroom door to whisper. 'Yes, it's me. Don't wake dad.'

'Did you have a good time?'

'Yes. Lovely.'

'What time did you get back to Barbara's?'

Here it was: the lie! Not too late to cause alarm, not too early to be believed.

'Half-past one. We just saw the New Year in and came away. But I couldn't sleep much. You know, strange bed. So I think I'll have an hour now.' An hour! She could sleep for a week.

'Don't you want some breakfast?'

'No thanks.' The drinks of the night before sat uneasily on her stomach. 'Just some sleep.'

'All right, get to bed. Don't sleep on your electric blanket.'

'I won't. Happy New Year, mum.'

Jenny switched on the electric blanket while she shed her clothes but the bed was still icy when she stepped into it. She tried hard to stay awake to give the blanket time to work but it was useless. She put down a hand and turned it off, pulling up her knees and putting her toes against her thighs for warmth. Gradually the chill left the sheets and she thought of Jed, rising from the Rover like Neptune from the sea. Dear, funny Jed!

But before she slept it was Keir's face that filled her thoughts – dark beneath the blond thatch, with straight thick brows above turquoise eyes. He had hardly looked at her tonight, except at midnight when he had kissed her as he kissed every other girl in the room. Still, there was time. When she came down from Liverpool there would be more opportunity. Somehow, some way, she would make him really look at her. 'This is my year,' she thought and, comforted, drifted into sleep.

May Denton was awake when her son Alan came into the house. She heard him fill the kettle and let out the cat, and then the soft plop as the gas ignited on the back hotplate. She listened as he moved around, assembling tea things on

15

a tray; she heard tea being spooned from the caddy, heard it scalded by boiling water and then the grating noise of the tea-pot lid as it was slipped into its groove. She could hear everything that happened in this house. It was her shell, the only place in which her heartbeat was steady, her mind serene.

'Mother?' His voice from the other side of the door was hushed in case she was asleep.

'Come in, Alan.'

He put down the tray and bent to kiss her cheek. 'Happy New Year.'

'Did you have a good time?'

'Yes.' He was reaching for her bed-jacket, holding it for her arms. She shivered in spite of her long-sleeved night-gown and he frowned.

'It's cold up here. Drink your tea.'

The china on the tray was Copeland and had belonged to Alan's grandmother. It was patterned with blue daisies, and May had embroidered the traycloth beneath it in matching colours. She liked things nice.

'Do you want toast?' He grinned suddenly. 'I could do you a fry-up, if you like? Start the New Year in style.'

There were only two links of sausage in the refrigerator and two rashers of bacon – enough for Alan today and tomorrow. On Wednesday she would have her widow's pension.

'No, thank you, dear. I never eat breakfast, you know that.'

He was hesitating, almost as though he wanted to sit down on the edge of her bed and talk. May felt a sudden flutter of panic and raised her cup to her lips to cover it. 'Run along to bed, now, and get some sleep. I'll hold lunch back till one.'

Did he look relieved as he straightened and made for the door? If he did, he only mirrored her own emotion.

Kath waved as Euan's Rover swerved down the cobbled back street of Belgate. The yard doors were all shut and the windows eyeless and curtained – forty houses in the terrace, and not a sign of life.

16

She went up the yard and mounted the back steps. The door was unbolted, the kitchen a sea of unwashed glasses and dead bottles. One thing you could say for the Botcherby family, they always saw the New Year in in style. Two inert figures were slumped in the living-room armchairs, and another on the settee.

Kath inspected them and recognized one as her brother and the others as his mates. She regarded them as they lay, lost to the world, mouths open, sleeping off the excesses of the night before. Drunken buggers! Still, they would be back to the pit tomorrow and not much rest after that. Let them at least enjoy the first day of a new decade.

She moved past them to the stairs that led up from the living-room and began to tiptoe to bed.

'Kath?' It was her mother on the landing. 'Happy New Year, pet. How many's down there?'

'Our Ted and his mates . . . Jim Skerry and Poulson's lad. Dead to the world.'

'I'd best get on, then. They'll be screaming for fried bread next. Don't disturb your dad, he's sound.'

'Want a hand?'

Her mother was rolling up the sleeves of her cardigan as she began to descend.

'No, pet. I'll eat that lot, once I've got hot water. Please God the fire's not out. Get yersel' to bed and get some shut-eye. You've still got those essays to do, mind on. It can't be one long holiday.'

Kath put out a restraining hand. 'I'll do them, you know that. I'm going to pass and make you proud of me. Remember the mink I'm going to buy you?'

'Never mind the mink, lady. You get a qualification and get yourself out of Belgate, that'll be my mink. Now, stop slavering and let me get on.'

'OK. But give us a kiss for Christmas.'

'It's not Christmas, daft ha'porth,' her mother said, but she let herself be kissed just the same. Kath was her only daughter, born after three boys to fulfil a longing for a girl, a girl who could be dressed up and reared without any thought of a future in the pit. She had longed for a pretty girl. That Kath had turned out to be clever too was an unexpected bonus and one to be exploited.

Euan and Jed dropped Keir at his corner, and turned on to Colliery Road. 'What do we do now?' Euan asked.

'Go home, I suppose,' Jed said mournfully.

'Your place?' Euan's tone was hopeful and Jed grinned. Euan was not over-fond of his imposing family seat just over the Northumberland border. Jed had walked through the echoing rooms and he understood why. It was comfortable enough, but more like a ritzy hotel than a home.

'Our place, if you like. There's sure to be a few cans left.'

'I was thinking more in terms of tea and toast.'

'Tea and toast.' Jed was mimicking Euan's precise speech. 'I think the mater can manage that. Tell you what, though, I know where there might still be some activity . . .'

'A party?'

'What passes for a party – music, booze, female companionship.'

'In Belgate?'

'In Belgate! Down there, left . . . no, right. Right, silly sod.' The Rover halted, trembled and swung right.

'It's eight o'clock, Jed. Surely they'll all be in bed?'

'Not colliery folk, my son. We've got stamina. No blue blood, you see. All solid red stuff, built on generations of brown ale and coal dust.'

'You're sure?' Euan's tone was doubtful.

'Bloody sure. We haven't even begun yet . . . hear that?' The throbbing music could be heard even inside the car as they drew up outside the terraced house and summoned up their second wind.

Keir let himself into the yard, hearing the soft cooing from the shuttered cree. So his father was still in bed. Usually at eight o'clock he could be found in the yard stroking a favourite bird, culling a shirker, handing out corn from a hand the size of a shovel. 'They get all his tenderness,' Keir thought. 'At least, they get more than me.'

He knew what emotions he aroused in his father: pride, run a close second by ambition. Pride in a son who had escaped the pit for Oxford, ambition for a son to live out his own fantasy of Labour rule. That ambition had been there from the beginning, from the day they had named him Keir Hardie after a socialist idol and ensured he would be ridiculed at school.

He looked around the neat kitchen with its apple-green paint and mock-tile wallpaper. If he stretched out his arms he could touch both walls. How did they stand it, and not only stand it but think it was the height of bloody luxury?

Keir spooned tea into a mug and scalded it, skimming stray leaves from the top before adding milk. He was just about to drink when he heard his father's stockinged feet on the stairs.

'Aye, you're back then.'

'Happy New Year, dad.'

'Not much chance of that with the Tories in. Still . . . have you mashed tea?'

'No, just a mug. I'll make you some.'

'Go on, then. Had a good night?'

'Yes.' As Keir spooned tea into the pot he wondered how much to tell. 'We stayed in Sunderland . . . friends' houses mostly . . . it was OK.'

'How did you get back? Shanks's pony?'

Dangerous to say 'yes': he might have been seen alighting from the car. In any case, why should he lie?

'I got a lift from Euan Craxford.'

His father slurped his tea and then shook his head. 'You'll get nowt from that direction.'

'He's OK. There's no harm in him.'

'He's a bloody Tory, isn't he?'

'I neither know nor care, dad. And we've had this out before. I have to mix with people, all sorts of people. I can't pick and choose by their politics.'

'Mix with them . . . mix with them by all means . . . you have to *mix* with them but you don't have to like them. That's when you go soft. I saw it in the old days, when the Londonderrys ruled the bloody roost. "No harm in the old Marquis," they used to say. "Canny chap, the Marquis." They were still saying it when he took the bloody bread

19

out of their mouths. They do it now with the managers: "He's all right," they say. You know what I tell them? He's a bloody gaffer's man, I say, so he can't be all right.'

Keir switched on the radio, hoping for diversion, but there was only more news of Afghanistan and the impending steel strike.

'Mark that, lad! The first national steel strike since '26. She's been in eight months . . . eight short months . . . and already she's trying to pick off the unions.' His father's final words were lost on Keir as he mounted the stairs to bed.

When his son had passed from sight Steven Lockyer went to the back door and gazed out on the winter sky. It had been a risk, letting the boy go to Oxford: you could get tainted in that kind of situation. Still, he had good socialist blood on both sides. And like it or lump it, you needed an education if you were going to achieve things. That was the way of the world.

The pigeons, sensing his presence, became agitated behind the green louvres. He reached for his boots and walked gingerly over the frozen yard.

The first bird rose up into the air and hovered above his head for a moment, its wings a blur so rapid was the movement. 'By God, you're bonny,' Steven told it, feeling emotion surge at the familiar sight. Not even Thatcher's lot could come between a man and his birds. He plunged chilly fingers into the seed and began to indulge his favourites, hunching his shoulders against the icy cold, feeling the muscles contract and swell. He could still hew with the best of them, fifty-five or not.

But his satisfaction was short lived. When the Belgate pit closed he would have to take redundancy, the fatal handshake. It was that or be bussed to another colliery, and he wasn't having that. He was someone in Belgate, his name a byword: Compen. Secretary for twenty-three years and a union stalwart. He'd stood up to more managers in his time than he'd had hot dinners. He would be nothing in another pit, a stranger in an alien land, too old to start again. So he would take the redundancy money and sell his job when the time came, learning to despise himself in the process. Not that all was lost yet: they must fight the closure, fight hard. He was still arguing against it, in spite of greed in the eyes of men promised money for relinquishing the pit. Never to go down the hole again *and* have a pocketful of money – that was how they saw it. It would take all Steven's skill to swing them his way.

He held out corn to the old cockbird. It would have to go soon; its wind was gone and it was useless, like he would be when he packed in, neither use nor mak to anyone.

He was feeling despair until he remembered his son. Keir! Keir Hardie Lockyer, a name to be reckoned with. He had prayed for a son to any and every god he could put a name to. The sods had kept him waiting fifteen years, until he had thought Cissie barren. And then they had relented and given him a cracker.

Steven smiled, thinking of his son's achievements and of what was still to come, more glorious than anything that had gone before. It was a new decade, after all. Another chance for everyone.

2

15 January 1980

'Jenny!' Jenny turned at the sound of her name, scanning the crowd in the market square. A thin rain was falling and people hurried past, heads down, moving towards the shelter of the station.

'Jenny . . . it's me.' Kath's brown hair was plastered to her head but the face beneath was cheerful. 'God, I'm soaked. Got time for a coffee?'

They darted into the snack bar and slid into a booth. 'I'll get them,' Jenny said. 'It's my turn.'

The coffee was weak but wonderfully hot and they held it in cupped hands while warmth crept back into fingers and toes.

'I hope it fairs up before tonight,' Kath said, squinting upwards at her wet hair.

'What'll you do with it?' Jenny asked.

'Tong it, I suppose. I won't have time to wash it. It'll be five before I get back to Belgate, and we're catching the six-fifteen bus.'

'Isn't Euan picking you up?'

'He offered but we said no. Can't make use of a bloke on his last night.'

'I still can't believe he's going,' Jenny said. Tomorrow Euan was leaving for Rhodesia, to work with his uncle on a tobacco farm outside the capital. The news had come as a bombshell to all of them, especially to Euan himself.

'Why doesn't he say no?' Kath said. 'He's twenty-three – his family can't make him go.'

'They can, in a way.' Jenny loosened the neck of her anorak and slipped her arms out of the sleeves. 'He's not like the rest of us, Kath, used to living on a shoestring. How would he manage without a car, for one thing?'

'Badly,' Kath admitted. 'Still, he'll hate Rhodesia – the part he'll see, anyway. All white man's rule and dinner with the governor.'

'He's seen it, remember? He was out there for a while last year. Things are changing, too. Once they have an election there'll be a difference.'

'I'll believe that when I see it. Still, I'll miss the guy. I couldn't stand him at the beginning, he made me cringe. I just thought he was some upper-class twit Jed had picked up. I mean, he is like something out of Jeeves! Now, though, he's one of us. It's funny, because we've only really known him for a year.'

'Two years,' Jenny said. 'He turned up with Jed one night in the Christmas holidays. I remember because it was the week we got Sooty.'

'Time flies,' Kath said. She was rummaging in the dress bag at her side. 'I got this for tonight.'

'Nice,' Jenny said. 'What'll you wear it with . . . your black skirt?'

'Oh, I don't know,' Kath said airily. 'There's so much else to choose from.'

Jenny nodded in sympathy. 'When I've got a job I'm going to have a wardrobe the full length of the room. It'll take me half an hour to decide what to wear.'

'No, you won't,' Kath said decisively. 'You'll have two good suits, nice little white blouses with real gold jewellery and a *huge* handbag with everything in it for every eventuality. And I do mean *every*.'

'OK, cleversides. I know I'm a bit fussy . . .'

'No, you're not.' Kath put out an affectionate hand. 'I'm only kidding. You *will* be organized but you won't be obsessive about it. Barbara's the one who'll be obsessive.'

'She makes lists of everything. Everything!'

'Don't tell me, I only share a room with her for half the year. Even the lecturers make jokes about it. "Have you got that down, Barbara?" they say. And she doesn't twig they're scoffing her. "Yes, thank you," she says. She has a hide like a rhino.'

'Still,' Jenny said, the old guilt overtaking her, 'she's OK really.' Barbara had been her friend for sixteen years – impossible to desert her now.

'Well, she brought us together,' Kath said. 'Without Barbara I'd never have met you and that, my girl, would have been tragic.' She rolled her eyes and raised both hands aloft in mock-horror.

'Everyone's looking,' Jenny said, grinning.

'Good,' Kath said, 'I like an audience. Now, get down to nitty-gritty. We'll get into Durham on the bus at twenty-five to. Are you coming with the Ba-lamb?'

'Yes. And Alan's manoeuvred her into picking up Elaine.'

'She's a cow about Elaine, isn't she? God, I can hardly wait for the summer. I'd never have gone in for teacher-training if I'd known I'd have to room-share with Babs.'

'I wish I hadn't promised to go on holiday with her,' Jenny said ruefully.

'Pull out,' Kath said.

'I can't. We're staying with her aunt and uncle. They have a guest-house in St Helier and she keeps telling me they've turned down bookings to take us. It's meant to be our big treat for getting through exams.'

'Some treat. Still . . . Jersey in June or July – it can't be altogether bad. You can lie on the beach all day and attract a millionaire. That's what it is, you know, a millionaire's playground.'

'More to the acre than anywhere else in the UK,' Jenny said wearily. 'Barbara's told me that *ad nauseam*.'

'Has she got them listed?' Kath said, straight-faced.

'She's working on it,' Jenny said. 'In descending order of affluence.'

'We're kidding,' Kath said, as they got up to go, 'but I wouldn't put it past her.'

Outside the rain had ceased but the sky was dark and lowering. 'I can't be late out tonight,' Jenny said. 'I've got my interview tomorrow, remember?'

'God, yes. I'd forgotten that. Our Jenny going job-hunting six months in advance . . . I said you were organized. I don't suppose we'll be late tonight, anyway, because Euan's flight is at the crack of dawn.'

They parted at the station, Jenny to finish her shopping,

Kath to catch her bus. 'See you,' Kath said and then, as the rain began again, 'Oh shit.'

The café where Elaine worked in the daytime was half-empty. Jenny found Elaine at the back of the shop, filling plastic tomatoes with ketchup.

'Ugh,' she said. 'That's gooey.'

'It tastes awful,' Elaine said, 'but the customers lash it on everything. 'Cos it's free, I suppose.'

'I just wanted to check on tonight. We'll pick you up about half-past six, is that OK?'

'Yeah. I'll come to the junction and keep Barbara happy.' They both grinned. 'Why are we going to Durham?' Elaine continued. 'It seems a long way for a few hours. Jed says we'll be finishing early 'cos of Euan's flight.'

'And because of me,' Jenny said. 'I'm off to Coatham first thing. I don't know why we're going to Durham, to be honest. Some idea of Euan's, I think.'

They chatted for a moment or two until Jenny felt the proprietor's reproving eye on her.

'I'd better go.'

'Take no notice of him, he's always like that. Smiles cost money, according to him.'

There were hollows in Elaine's cheeks and tiny blue veins in her eyelids. Jenny drew in her own rounded cheeks, longing to have Elaine's ethereal look, but even as she did it she knew it was a waste of time.

'I've got to go anyway. See you tonight.'

It was growing dark outside and Jenny was hurrying towards the bus stop when she saw Keir standing on the bank corner, talking to two strange girls. She watched as he bade them goodbye, seemingly unaware of their adoring expressions, and moved off, swinging his scarf around his neck. On an impulse Jenny turned in her tracks and crossed the road towards him. She kept her eyes averted, as though she had not seen him, and waited until he saw her.

'Jen! What are you doing here?'

'Shopping . . . and I've just popped in to see Elaine.'

'She OK for tonight?'

'Yes, we're picking her up about six-thirty.' Keir was

relieving her of her parcels and taking her elbow to guide her through the crowds, and Jenny tried hard to think of something intelligent to say.

'It'll be funny without Euan, once he's gone.'

'We'll miss his car.' Keir was looking down at her and she felt her cheeks flush. 'Don't look so shocked, Jenny, I'm only kidding. Of course we'll miss him, he's a good lad. And I'm sorry for anyone going out to Zimbabwe now . . . his type anyway. They've had it.'

She might have known he would say 'Zimbabwe'. To everyone else it was still Rhodesia, would be for a long time, but Keir understood these things.

'This your stop?' She nodded, still tongue-tied. It was different when they were all together, when conversation flowed and every second word was a gag. Being alone with him was hard. Perhaps she had wanted it for too long and all she could do now was to make banal small talk.

It was almost a relief when her bus came into sight and she could wave to him through the glass until he was lost to view.

On the stroke of four May Denton drew the curtains, moving from room to room, patting a cushion into place here, flicking an imaginary speck of dust there.

In Alan's room she paused when the curtains were closed and gazed around her. He was neat as his father had been. She closed her eyes, struggling to summon up her dead husband's face. He had had an Anthony Eden moustache and he had been such a gentleman.

Suddenly she remembered his lips against her cheek, the moustache tickling so that she shrank away. He had only kissed her once upon the mouth. She had not liked it and he had never pressed her again. Afterwards she had wondered . . . but he had always told her she was a good wife. 'Splendid,' he had called her once. 'Splendid.' And she had kept her word to him, over all these years. He had wanted a son with a profession and soon he would have his desire.

She put her hand to her mouth to curb a smile of satisfaction. Never count your chickens. There were still the final exams and then all the complicated ritual of being accepted at the Bar. That cost money

Thinking of money reminded her of the immersion heater. Alan was going out tonight, he would need hot water. She went downstairs and put down the switch. Half an hour would be enough to warm it up. She moved into the kitchen and reached for her pinafore, tying it around her waist and then making to draw the curtains.

Across the street the Wilsons' lights were blazing in every window. Apart from the waste, it showed them up like figures on a cinema screen, larger than life. She twitched her curtains on the sight of husband and wife moving together. It felt better with the curtains drawn. Safer, somehow – and besides, it kept in the heat.

The cod she had boiled at lunchtime was cool now, resting in a sea of scummy water. At her feet the cat began to purr and weave. 'Patience, patience,' May said, lifting the fish on to a plate. The skin came away cleanly and she carried it to the cat's dish.

'There now,' she said, 'don't make a mess.' She flaked the rest of the fish carefully, seeking out bones. The pan of mashed potato was still warm as she added salt and pepper and a pinch of dried parsley. She stirred in the fish, trying not to break up the flakes. It was so much nicer if you could identify the fish.

When the mixture was ready she formed it into cakes, three for Alan and one for herself, dipping them in seasoned flour ready to fry. When they were done she began to peel sprouts, scoring the base with a cross and counting them into the colander. Eight for Alan and four for her.

There was still rice pudding in the larder and she added raisins before putting it over a pan of water, ready to heat through. He could eat as soon as he came in, and then get ready.

Suddenly she thought about having the house to herself tonight, only her and Tiddles and her radio. She would have her wash-down early, and get all ready for bed before the Music Programme. As if it heard and agreed the cat miaowed at her feet. May looked at the clock. 'Not long now,' she said. 'Not long now.'

She fried the fishcakes carefully, her mind drifting as it

27

often did nowadays to the past. Alan had always been a good boy. Even as a baby he had never demanded attention. It was as though he knew about the gossip, the scathing glances when she took him out in his pram. Knew that he must be on his best behaviour for her sake.

She smiled, remembering Alan in his red siren-suit, toddling in front of her on the reins, always anxious to get somewhere. She had known he was different, then. That had made it possible to bear the taunts, the stares, the hidden smirks. That was the picture that always came when she thought of him: the little boy in the red suit in the park on the day she found pride.

When she finished the frying she looked with satisfaction at the assembled meal. It looked appetizing and it hadn't cost too much.

She had grown up in a home where money had been scarce, and then had gone into service at Longden Hall, where thrift in the kitchen was obligatory. How often she had blessed that early training. Without it she could never have managed. She draped the dishcloth and hung up the tea-towel and then took down her jar of Marmite. Just enough water in the kettle to dissolve a spoonful. She cupped her hands around the warm mug and sat down at the table to drink.

She had stayed at Longden Hall for six years, until she was twenty. She had watched and learned, and never forgotten her mother's advice not to meet the mistress's eyes: 'It makes them think you're bold, May, and they don't like that.'

The first time she had looked the mistress in the eye was the day she gave notice. 'I'm going into munitions. I'm sorry to go but I have no choice.'

She had lived in Birmingham for the rest of the war, not drinking or smoking like other girls, saving her money and planning for the day peace came when she could go anywhere she pleased.

That was a long time ago. A lifetime.

Jed was already waiting when Kath reached the bus-stop, and Keir was just behind her.

'I'm not used to this,' Jed said gloomily.

'Used to what?'

'Bussing it.'

'Come off it,' Keir said. 'You've been bussing all your life . . . or legging it.'

'Not lately. That Rover's been a godsend this holly. Still, poor old Euan, it'll be worse for him.'

'Then why's he going?' Kath said in exasperation. 'He could tell his family to bugger off, and get a job here.'

'They want him out of sight,' Jed said. 'He doesn't fit the pattern. He's not sporty, he doesn't swan around with the upper crust, and he's flunked his university place. He drinks too much and plays jazz. Ergo . . . they want to lose him.'

'He's so canny,' Kath said.

'We know that,' Jed said. 'We have discernment. But his lot have other values.'

Keir made some reply but Kath was no longer giving the conversation her full attention. On the other side of the road David Gates was jogging along the dark gutter, against the direction of traffic. She could recognize him from the fair head above the red track suit.

'There's Davy Gates,' Jed said.

Kath nodded. 'I was at school with him . . . same class.' She felt a flutter of excitement as the jogger drew near.

'He's had a try-out for Hartlepool,' Keir said. 'Dad told me. Now he plays for the Belgate Stars.'

'He's still at the pit,' Jed said. And then, as the boy drew level, 'Aye, aye Dave. Still at it, then?'

The jogger nodded and smiled without breaking the rhythm of his step. 'Still at it, Jed. Keir. Still a bonny lass, then, Kath.'

He was past now and Kath's cheeks were burning.

'Well, well,' Keir said. 'So you've got an admirer, Kath. Does he need an eye-test?'

'Very funny, Keir.'

'We'll be bridesmaids,' Jed offered, pulling his long green coat around him and arching his back. He moved forward out of the bus shelter, humming Mendelssohn and holding an imaginary train.

'Can it,' Kath said, and then, before she could reach out to thump him, Euan's Rover was drawing alongside.

'I thought I might as well,' he said mournfully, leaning to open the passenger door. 'I had bags of time. Hop in and let's get on with the wake.'

'And don't be late, Jenny. Tomorrow is important.' Her mother's words were ringing in Jenny's ears as she ran down the path to Barbara's car. Alan was in the front seat and Jenny slipped in behind.

'I hope she'll be ready,' Barbara said. They both knew she meant Elaine and Jenny sought a glimpse of Alan in the rear-view mirror. He was smiling but he didn't look amused. Perhaps he was keen on Elaine? He had always been good at hiding his feelings, right from the first time Jenny had seen him in the school yard, a big ten-year-old to her small eight.

Elaine was waiting on the corner when they got there, shivering inside her thin black cloth coat, the long beautiful legs looking slimmer than ever in black nylon.

'You look freezing,' Jenny said, when she was safe in the back. She reached out for one of Elaine's hands, holding it between her own warm palms, seeing where the nails were bitten to the quick and sore around the cuticle. The hand was long and thin and would have been elegant except for the bitten nails.

'Ta,' Elaine said when Jenny let go. 'Where are we going?'

'Durham,' Barbara said shortly.

'We're going to some Indian restaurant Euan knows,' Alan explained. 'He's paying. Then we're going on to a disco – not that there'll be much time.'

Beside Jenny Elaine shivered, this time with anticipation. 'It sounds all right. I'll miss Euan, though. I wish he didn't have to go.'

'Perhaps he'll become a tobacco farmer and have a plantation,' Barbara said.

'And we can all go and stay there for long, long holidays,' Alan finished for her, turning to wink at Jenny.

'Well, it would be marvellous,' Barbara said. 'They live

like lords out there. You just clap your hands for servants, and they dress for dinner.'

'I thought they wore grass skirts and had bones in their noses,' Jenny said.

'You know what I mean,' Barbara said scornfully. 'I didn't mean Africans, I meant people like us.'

And then Durham cathedral was there in the distance, floodlit and magnificent, the castle squat and brooding below it.

'It gets me every time I get the train,' Alan said. 'I look at it and I get the old lump in the throat. It means home, I suppose.'

They ate in an Indian restaurant, dimly lit, darkly decorated. The waiters smiled a lot and sometimes looked anxious when their hilarity threatened to get out of hand.

'Euan will be back inside a month,' Jed predicted. 'You watch. He'll miss Britain, good old Britain.'

Barbara pounced. 'So you admit it's a good country, Jed! That makes a change.'

'The country's all right, it's the poxy government I can't stand. Still, Thatcher won't last. I give her two years.'

'We'll see,' Barbara said smugly. 'When people appreciate what she's doing they'll vote her in forever. Stick to your psychology, Jed and leave politics to the grown-ups.'

'God, you're dim, Barbara. For the tenth time I am not going to be a psychologist. If I was I wouldn't need to do five years of medicine. I am going to be a psychiatrist. P - s -y -c -h -i . . .'

'Lay off, Jed,' Alan said quietly. 'You and Barbara can duff one another up tomorrow. This is Euan's night, remember?'

Jed raised his glass in Barbara's direction. 'OK, Ba-ba-black sheep, tomorrow I'll try and put you straight. In the mean time, here's to luck. May it bring me a bloody good degree.'

'I'm trusting to hard work for my qualifications,' Barbara said. 'Some of us worry about the future.'

'We are the bloody future, darling,' Jed said. 'More's the pity in some cases.'

Euan was an expansive host, urging them to eat and

31

drink, laughing uproariously at every joke, but Jenny guessed he was unhappy.

'Here's to Euan,' Keir said when they reached the coffee stage. 'May he enjoy Africa, and come back soon.'

'Here's to Zimbabwe,' Jed said, when the first toast was drunk. 'May it survive having a loony dumped on it, and may it bloody well enjoy its independence.'

They drank to that with varying degrees of enthusiasm.

'They haven't got independence yet,' Barbara said.

'Oh, come on.' Jed was leaning across the table. 'Face the facts, Ba-ba, it's only a matter of time. They've bloody well struggled for it, don't begrudge it.'

'Oi!' From the end of the table Alan's glare was chilling.

'Sorry,' Jed said. And then, trying to scramble across the table . . . 'Kiss better, Babsy.'

'Come on.' Keir was hauling him back, reaching to the rack for Jed's bizarre, almost floor-length coat. 'Let's get out of here.' He turned back to Euan. 'Good grub, Euan. When you come back it'll be on me.'

'On the terrace of the House?' Euan said.

'Ministerial chambers, old boy,' Keir said mockingly, but Euan's face had clouded.

'You'll all go on and do things, and I won't be here.'

Alan waited at the door while Euan settled the bill. 'Why are you going?' he asked him bluntly as they emerged into the street.

'I don't have much choice. The family insist I do something. My uncle farms tobacco, and the intention is that I see if it interests me. I hate the whole idea, but what else can I do? I have two talents, Alan: I drive well, drunk or sober, and I play drums. Quite well, I think. Well, sometimes quite well. But there's no way I could hold down a job.'

'You never know,' Alan insisted. 'You've never had a chance to test yourself.'

Before Euan could answer there was a holler from the group in front. 'Get a flipping move on! We've only got another hour.'

They spent the last hour in a smoky disco, gyrating to music, swopping partners casually so that it was never

32

clear who was with whom, and gradually, as the minutes ticked away, turning their attention to Euan.

As they emerged into the street for final farewells they were suddenly subdued. 'Take care, mate,' Jed said, putting his arms around Euan and giving him a bear-hug.

'We'll write,' Barbara said, and Jenny squeezed Euan's hand and nodded, suddenly bright-eyed. Euan kissed Elaine and then the other girls, and was enfolded by each of them in turn.

'Go on, you daft brush. Go and find your car and scram,' Kath said, 'before we all blub. You know where we are when you're ready to come back.'

'See you,' Keir said.

'I'll set you to the car,' Alan said and fell into step with him. The others watched as Euan turned once and waved, and then they were gone.

'I'll get the car,' Barbara said. 'You stay here, Jenny, and wait for Alan.' Once more she was pretending Elaine did not exist and Jenny linked her arm firmly through Elaine's in a show of solidarity as Barbara moved off.

'We ain't got a lift no more,' Kath said, pulling a face.

'No.' Keir pushed his hands into his pockets. 'And the last bus leaves at eleven.'

'I'm not going back to Belgate,' Jed said, and they turned in surprise. 'I'm staying here in Durham with my Aunty Mamie. I promised Mam I'd see her before I went back to Edinburgh, and this is the only chance.'

'She couldn't put three up, could she?' Kath said hopefully. She was beginning to shiver as the cold struck home.

'Nee chance, pet. She's got a one-bedroomed Aged Miners' cottage. I'll be on the settee. She'll brew you a pot of tea, though.'

'We'd miss the bus,' Keir said hastily, shrugging into his coat.

If only Barbara were generous, Jenny thought, they could all squash into the Escort, she close enough to Keir to feel his heart beat. But even as she thought it she dismissed the idea: there were too many of them.

'Cheer up, Jenny. He'll be back.' Keir was regarding her ruefully and she smiled.

'I know. It's just that he's the first one of us to go.'

'Yeah,' Jed said. 'We haven't known him that long but he's one of us, all right. And I am off as well! I need a pee and it's Aunty Mamie's or the back street.' He put his arm around Elaine and kissed her brow. 'Bye-bye, sweetheart. When I'm rich and drive a Bentley you'll never need to go home with the Snow Queen again.'

He went off up the street, soberly at first and then, unable to contain himself, leaping on to a low wall to walk along, balancing himself with outstretched hands. At the end of the street he leaped into the air and landed, knees bent, turning as he straightened to give a last wave.

'Do you think he's all there?' Kath asked.

'No,' Keir said. 'That's why he'll make a good psychiatrist.' He linked his arm in Kath's. 'Come on then, Katherine, it's thee and me for the Belgate bus. Good luck with your interview tomorrow, Jenny. Don't worry, you'll walk it.'

'Yes,' Kath said. 'Tell them they'll be lucky to get you.'

When Kath and Keir were gone Jenny looked at Elaine. The girl's face was pale under the street lamps. 'You look frozen, Elaine.' There was a doorway behind them and they retreated out of the wind, peeping out now and again to look for the Escort.

'Barbara's taking a long time.'

'So's Alan.'

They huddled together for warmth, watching the night sky, frosted steel above them, the pavement glistening like marcasite. When they heard the car they stepped forward, but it was too late – the Escort had slowed and then, seeing an apparently empty street, Barbara had sped on.

'Stop,' Jenny shouted, but it was useless. The car turned at the corner, down towards the bridge.

'She's going the wrong way,' Elaine said.

'She's probably looking for us,' Jenny said despairingly. They stood on the pavement, disconsolate.

'What do we do?' Elaine said.

Jenny looked at her watch. 'We've five minutes to the last bus. I daren't risk missing it and then missing Barbara. You know her, she won't look for us for long.'

'But where's Alan?'

'I don't know, but he can take care of himself.'

They walked towards the bus station, heads jerking hopefully at the sound of every approaching car, dropping again when it turned out to be someone other than Barbara.

'It's all gone wrong, hasn't it?' Elaine said mournfully.

'Never mind,' Jenny said. 'There's the bus. We'll be OK.' But as she settled in her seat and turned to look out once more in case Alan or Barbara appeared, she felt that Elaine was more right than she knew. It was not just the night that had gone wrong, it was the charmed circle of their friendship that was broken now. She tried to compose herself as the bus moved off and think of the interview ahead of her, but it was the image of comrades gradually drifting apart that filled her head.

3

16 January 1980

It was five past twelve when Jenny reached the *Coatham Gazette* building. While she waited outside the editor's office she fidgeted with her grey suit and her bag and gloves. Her mother's cameo brooch was pinned at the neck of her white shirt and she touched it every few minutes in case it had come loose.

'Sorry about the wait,' the editor said, popping his head out to summon her into his untidy sanctum. 'Sit down. Coffee?' He handed her a plastic cup and refilled his own. There were at least three dirty cups on his desk and an overflowing ashtray. He caught her tiny grimace of distaste and grinned.

'It isn't Fleet Street, is it? But this is the best place in the world to learn the trade. You get everything in a provincial paper, life, death, altruism and corruption. You name it, it happens in a small town.'

Jenny nodded and grinned in return. 'Yes, I understand. It's just . . .'

He threw back his head and laughed. 'You didn't think editors did anything? Well, on the *Coatham Gazette* the editor does everything. I even make my own coffee, which accounts for how lousy it tastes. Now, tell me, how are we related?'

'Your mother and my grandmother were cousins,' Jenny said. 'It's a bit distant, isn't it? I told them not to pester you, but you know what mothers are like.'

'Indeed I do. Still, I didn't suggest you should come and see me to please my mother . . . or yours for that matter. I liked what you sent me. Typical college stuff, but punchy. You can write. I wanted to have a look at you to make sure you didn't have two heads.'

Inside Jenny pleasure bubbled: he was going to give her a job! His next words brought her down to earth.

'I'm not going to take you on . . . not in the usual way . . . because I don't think it would be fair. Normally, you'd be tied to me for a number of years, and I think you could do better than that. But a period here would do you good. As it happens, my Women's Editor . . . grand title but she's really just another journalist . . . has got herself married. Her husband has to spend some time in America later this year, so she goes on leave at the end of August and hopefully she'll be back when her husband returns.' He reached for a pack of cigarettes, offered one to Jenny and when she refused lit up his own.

'You can't take her place, of course, but if you care to come in as a dogsbody, union rates, we'll teach you the basics. Perhaps let you try your hand at features. And of course if something else turns up that you fancy, the usual four weeks' notice. I'd try the BBC graduate entrance scheme if I were you. How does that sound – the job, I mean?'

'It sounds . . . fabulous.'

He cut short Jenny's thanks, apologized for not being able to take her to lunch, and escorted her to the door. 'Let me have a note of your expenses.' She could hear the hum of presses and there was a constant sound of telephone bells.

'Keep in touch,' he said and waved before vanishing through a door marked 'Advertising'.

Jenny had to restrain herself from skipping up the street. He had taken it for granted that she would graduate, and he liked her work! There was a train back to London at two-forty, and if she caught that she could be in Sunderland by eight o'clock, or shortly after. There was sure to be someone in the coffee bar . . . maybe even Keir. Perhaps tonight would be the night, the special night?

She bought a bar of fruit-and-nut from a vending machine and ate it before ringing home.

'Mum? Yes, fine. Yes, very nice. Yes, he did. To start when I leave Liverpool. Just for six months or so but that's partly for my benefit. No, I'll explain when I see you.' At the other end of the line her mother chattered anxiously.

'I don't know which train. Tell dad not to worry, I think Barbara's coming to meet me.' She crossed her fingers at the lie before she went on. 'I don't know what time I'll be in . . . not late. Promise. 'Bye.' And then, suddenly guilty . . . 'I love you.'

All through the long journey home Jenny hoped Keir would be in the coffee bar – and when she got there he was. But one look told her there would be no ecstatic tête-à-tête. He looked strangely gaunt, as though he had aged, and for once the sleek blond hair was tousled.

'Hi . . . what's the matter? You look . . . rough.'

Jenny was sliding into the seat until the dumb misery on Keir's face stayed her.

'What is it, Keir? What's the matter?' It must be his mother or father, she thought wildly. Nothing else could make him look like this.

His eyes met hers, screwed up as though he was in pain.

'It's Jed. He's dead.'

Jenny sat down on the leatherette bench, bewildered, her ears singing, a sudden nausea rising in her throat.

'Don't fool, Keir. It's not funny.'

'I'm not fooling, Jenny. I wish to God I was. I keep telling myself over and over that it's not true. Not Jed. I can't take it in. But it is true.'

'How? Was it a car?' The picture of Jed rising naked and smiling from the sun-roof came into Jenny's mind. 'When did it happen?'

Around them the half-empty coffee bar was suddenly throbbing with noise as someone fed the juke-box.

'Let's get out of here,' Keir said and rose from his seat like an old and drunken man. Jenny put out a hand to steady him and then unaccountably lost her own balance and felt a table edge harsh against her hip.

'How did it *happen*?' she asked again, when they were out in the deserted High Street where shop windows were peopled with vacant-eyed models already dressed in spring fashions.

'He fell into the river,' Keir said, shaking his head. 'No one knows how.'

'Oh God, I don't believe it.' It was her own voice, rising now to an unfamiliar shriek.

'Come in here,' Keir said and urged her into the welcome darkness of an arcade. He put his arms around her, holding her close. Jenny had longed for such a thing to happen, but now there was no feeling except grief.

'Tell me,' she said, her lips against his shoulder. 'Tell me everything you know.' In the street outside a police siren sounded, growing louder and then dwindling away.

'I know hardly anything,' Keir said. 'I went to his place as soon as I heard, but they're distracted . . . and they don't know much, only that he drowned. Last night.'

'*After he left us?*'

'Yes. He went down to the river bank – don't ask me why, but he did. A passer-by saw him there. He was with someone else and . . . he fell in.'

'But how?' And why did he go there? He was going to his aunt's.'

'That's where he *said* he was going,' Keir answered. 'He might've changed his mind, or he may never have intended to go to his aunt's. Maybe he was meeting someone from the disco? He would hardly have been able to say anything in front of Elaine.'

'Jed wasn't like that.'

'You can't be sure.'

'Was it a girl . . . the person who was with him?'

'I don't know, Jen. We might find out more tomorrow . . . and it'll all come out at the inquest.'

Inquest! One of those words you never, ever associated with your own life. Or with your friends. Jenny felt laughter bubbling up inside her and it wasn't a pleasant feeling. She put up a hand to her mouth and heard the crying start inside her chest.

'Now, Jenny, calm down.' And then more sharply: 'Cut it out, Jen.'

They moved eventually, out into the windy street where lamps seemed to shudder and litter blew like butterflies from gutter to gutter.

'I don't want to go home,' she said.

'No,' Keir answered and held her more firmly until they reached the park. The lake gleamed in the lamplight; the trees beyond it were a black chasm.

'We'll never see Jed again,' Jenny said – and then, feeling

rather than hearing Keir sob, she murmured, 'There, there' and began to comfort him.

They found a spot out of the wind, with trees on two sides, and a wall behind them. It had started to rain, spattering the trees, but she didn't mind. She kissed Keir kindly, rather as her mother kissed her when life was cruel. He kissed her back, his mouth gentle, a little moan escaping him now and then because they were there not in joy but in death.

'Hold me, Jenny.' She put her arms around his chest and squeezed him to her.

'Jenny, Jenny.' His lips were in her hair and on her cheek, soft and sad, and she twisted her head around so that their mouths could meet. She felt her thighs begin to tremble uncontrollably and then he was urging her down, his hands on her shoulders. She felt wet leaves stir beneath her knees, and then they were lying together, suddenly comfortable and relieved.

There was a long pause between grief and love, but when it came it seemed the most proper thing in the world.

'Is it all right?' Keir whispered against her cheek as he touched her breasts, and she shook him slightly, angry that he needed to ask. She felt him fumble at her waist and then move to the hem of her skirt. He would never manage on his own. She it was who pulled and tugged to make the way clear for him, to let his fingers find the way – and then a short, twanging pain as he entered her. Her mind split then, one half marvelling at how simple, almost commonplace, it was after all the years of waiting, the other half rejoicing that there was something she could do to bring Keir comfort.

'I love you,' she said as he moved inside her, glorying in the pain that was sharp enough to be significant but not too much to bear.

'I love you too, Jenny. I love you, love you, love you.' He was moving faster and faster and for a second she was afraid, realizing that she could not draw back. And then he moaned and she felt his full weight on her body.

'There now,' she said. 'There now.'

'I didn't know you hadn't, well . . . before,' he said and held her close again.

Afterwards they both cried and Jenny reached for tissues from her bag and shared them out.

'We're not crying for Jed,' Keir said, his cheek against her hair, his voice shaking. 'We're crying for what life's going to be like without him.'

'We'll have each other,' she said.

'Of course we will,' he answered and lifted her to her feet.

They walked back through the dripping trees, clinging to one another like drowning victims, stumbling on to the path and turning towards the lights of the road outside.

'Thank you, Jenny,' Keir said once and she shushed him to show it didn't matter.

He insisted on putting her into a taxi at the station, scorning the bus. 'Don't argue, Jen. Not tonight,' he said and paid the driver in advance. 'I'll see you tomorrow – I'll ring Alan. We've got to get together and find out what happened.' And then, almost as an afterthought: 'You're a very special girl, Jen. I think I might love you.'

Jenny wanted to ponder Keir's words in the taxi but all she could think of was Jed. If someone had killed him . . .?

And then suddenly she remembered her parents. How could she face them now? They had always made her promise to stay a virgin. They called it being 'good' or 'straight' but all three of them knew what that meant. And now she had crossed the Rubicon and there could be no going back. She tried to analyse her feelings, to see if there were any identifiable differences between the old Jenny of this morning and the new Jenny of this evening. Finding none, she concluded that virginity or the lack of it was insignificant compared to the loss of a friend. And knowing how it had come about, surely even her parents would understand?

'You know then?' her mother murmured as she entered the hall, seeing Jenny's tear-stained face. There was a poinsettia on the walnut side-table, its leaves the colour of blood.

'Barbara's been on the phone,' her father said, looking sheepishly sad for her. 'You'd better give her a ring, she's in a bad way.'

Jenny's fingers fumbled with the dial until her father

took the phone from her and held it until Barbara came on the line.

'I was so rotten to him, Jen. I didn't mean it, he just used to rile me . . . Now I feel awful. I can't even say I'm sorry, I can't ever make it up to him . . .'

'It was just fun, Barbara – Jed knew that.' Jenny didn't want to console Barbara, she wanted someone to comfort her. She kept thinking of Jed, joking, grinning, calling her Jenny Wren, folding Elaine into his long green coat . . . oh God, Elaine! She had been so taken up with her own feelings she had forgotten about Elaine! She sought escape in tears and cried down the phone to answering sobs until her father put an end to the conversation.

'Come on, darling, there's a good girl. Say goodnight to Barbara and get to bed . . . Mother's making cocoa. Get a good sleep and then see what's to be done in the morning.'

'What's to be done?' Jenny was looking at him stupidly – what could she do for Jed now? He *couldn't* mean what had happened with Keir? Before her father could answer there was a ring at the bell and her mother appeared, looking flustered, the mug of cocoa in her hand.

'Who can it be at this time of night?'

It was Kath, muffled up to her ears, her eyes red with weeping. Behind her, at the kerb, a tall boy sat astride a motor-bike.

'Kath!'

'Oh God, Jenny, I can't believe it.' Jenny moved towards her and they clung together until Kathy detached herself.

'I can't come in.' She nodded towards the boy on the bike. 'I can't be long. Dave only brought me as a favour, because I just had to talk to you.' She suddenly held Jenny at arm's length. 'Are you OK kid? You look awful.'

Jenny was unwilling to meet Kath's eye, remembering Keir and what had happened an hour ago. She didn't want Kath to know, not yet.

'I'm all right, I'm just numb. I can't believe it.'

'It's true. Jed was larking about on the river – with someone, they don't know who.'

'I keep thinking about Elaine. Has someone told her?'

'Yeah, Alan. Keir came through to Sunderland and told him when we found out this morning. Alan went and told

Elaine. She was shattered, but Alan stayed with her till a
neighbour came in. We're all going to meet tomorrow . . .
Luigi's at twelve. I've got to go now, I can't keep Dave
waiting. Get some sleep.'

'I don't feel as though I'll ever sleep again.'

'Me too. But we will, Jen. We'll go on. Jed would've been
the first to say we must.'

'We'll have to write to Euan . . .'

Kath shook her head. 'I've already phoned his mother.
She was quite nice, by the way. Not snobby. She's going to
tell Euan on the telephone.' She frowned. 'Don't look so
dumbstruck, Jen. Euan has a right to know straight away.
He was fond of Jed.'

'You're right,' Jenny said. 'I should've thought of that.'

She waited in the doorway until Kath had climbed on to
the pillion and bent herself to fit the driver's back, lifting a
hand to wave as they roared away.

'Come along,' her mother said then, and Jenny allowed
her father to usher her up the stairs.

'I know how you feel,' her mother said unexpectedly,
when they reached her room. 'There was a boy when I was
young – I liked him, but it was nothing special. He fell
under a tram-car. We couldn't get over it, any of us. Not
for a long time. You expect grandparents to die and nice
old ladies down the street. But not your equals, not
someone hardly started.'

Her mother was not usually given to reminiscence and
Jenny was grateful. She was hugging her when she caught
sight of her own face in the mirror, white and haunted,
and suddenly she was ashamed. Her mother still believed
in her, trusted her, was totally unaware of the uncomfortable wetness between her legs, the tiny pain when she
took a stride. Life was never going to be the same. Jed was
dead, and in dying he had changed the course of her life.

When her mother was gone she kneeled down and
prayed for Jed's parents and for Elaine and for what
happened between her and Keir to turn out for the best.
He couldn't think she was easy, not in the circumstances?
He would understand that it had been different, that she
had done it for a reason and not just for lust? She shivered

43

at the word and scurried for the bathroom. There hadn't been any lust in it, only love and the desire to give comfort.

She was thinking about Keir and love when she remembered what her mother had always said about 'consequences'. What if she became pregnant? In a funny way it would be all right – a life would have sprung out of Jed's death. All the same, it would be terrible.

When she took off her clothes she saw the red stain on knickers and petticoat, mute evidence of sin. She left them on the floor while she washed and then hid them in her dressing-table until tomorrow. In bed she pulled the blankets up over her ears, but she couldn't stop thinking of her father's kind face, and how it would crumple if he found she had let him down. She tried to switch her thoughts to tomorrow and what must be done. She would go and see Jed's parents: that was one thing she *could* do.

'Jed is dead,' she said in a whisper and then louder, like probing a mouth ulcer, *'Jed is dead!'* He had died in the swirling river, cold always, icy in January. He had died while someone stood on the bank and watched him drown. In her mind the figure on the bank took on a face, and the face was Alan's. She turned on her side, confused, and tried to sleep.

4

17 January 1980

Steven Lockyer walked past the Aged Miners' bungalows, marking where windows sparkled, where the odd pot of bulbs stood – any little sign of interest. He dreaded the dead windows, curtains limp and faded, like the occupants. You died in there. Quicker and easier to take to the waves and finish it rather than sit there, up above the seashore, waiting for the inevitable.

All the same, there was no certainty to life. Young Jed gone in a flash, hardly out of the pram. It was a fact, and no mistake. The thought that it might have been Keir, his own lad, suddenly struck him, so unthinkable that it had to be banished at once before it turned into naked fear.

'Aye, aye, Steven.' The man hailing him from the porch was grey of face and hair. 'Packed it in?'

'No such luck, Jack. I'm on tub-loading. Long time since I saw you. How long've you been down here?'

Jack Longman had been a deputy when Steven started at the pit, looking out for the boy, sampling his bait as reward, joining in the scoffing that was ritual but never letting it get out of hand. Now he was a shadow of the man he had been, only the big knuckles and broad wristbones to suggest he was once a giant.

'I came down here when the wife died. I didn't need the three bedrooms, not with the family gone. Your Cissie all right, then? And the lad?'

'He's at Oxford.' Steven tried to keep the pride from his voice, unsure how Jack Longman would take the news. Some folks were sniffy about others getting on. 'St Edmund Hall. He's studying politics . . .' Jack was looking suitably impressed and Steven expanded. 'You need it nowadays. It's not the workers' party any more, not for what I can see. So if you want a future you have to go with the stream.'

'He's got his eye on Parliament, then?'

'Aye, in the long run. He fancies a union job when he comes down from Oxford. They have all sorts of openings for likely lads. I've spoken to Durham . . . they're putting the word out.'

The wind was sharpening now, blowing in from the east. 'Fancy a brew?' the older man asked hopefully.

Steven pursed his lips and looked towards the sea, grey and forbidding now that the wind had changed. 'Might as well, Jack lad. Might as well.'

'Have you got the address?' Barbara asked, as Sunderland fell away and they were on the Belgate road.

'I know where Jed lived . . . we called for him once when I was in Euan's car. Just keep going till I tell you to turn.'

'I know he's dead,' Barbara said, 'but when we knock on the door – his door – I feel sure he'll open it and start fooling around.'

They were silent for a mile or so, each thinking of Jed, and then the outskirts of Belgate came into view: rows of terraced houses all leading towards the sea, the pit wheel standing sentinel above, and hardly a touch of green anywhere among the brick.

'It's awfully crowded,' Barbara said.

Jenny pursed her lips before replying. 'It was OK by Jed. He loved this place.'

'*Why* was he down by the river?' Barbara asked suddenly. 'And who was with him? He left us . . . no, Euan went first with Alan, and then I went to get the car. You and Jed were standing on the corner with Kath and Elaine and Keir. Who went next?'

'Jed did,' Jenny said, taken aback by the intensity of Barbara's questions. 'And then Keir and Kath went off for their bus, and Elaine and I hung on for you.'

'Not where I could see you. I was frantic, trying to find you . . . not to mention Alan. It was after eleven before I gave up.'

'Didn't Alan turn up either?' Jenny was surprised.

'No, he didn't; I couldn't find any of you. I drove round and round. It was spooky, hardly anyone about. And then I just gave up. I knew you must have managed somehow.'

'We got the last bus . . . just. I walked part of the way to Elaine's and then I got a taxi. I felt really good that night, really hopeful. I even thought Euan would come back soon.'

'Do you think he'll come back for the funeral?'

'It wouldn't make sense, not all that way – but with Euan you never know. Turn left at the next corner. It's about five doors down.'

'They've got no gardens,' Barbara said. 'Imagine having no greenery at all.'

They had no difficulty in finding the house. It was of red brick, impossibly narrow, with two windows above and one below, and a green door opening off the pavement. Its curtains were drawn and a little knot of neighbours stood across the way, their eyes on the façade that concealed bereavement.

'What do we say to them?' Barbara said, suddenly panic-stricken, while they waited for an answer to their knock.

'Just that we're sorry . . . and is there anything we can do, I suppose. Just play it by ear.' Jenny's voice sounded remarkably calm but her heart was thumping until the door opened and she forgot her own fears. Jed's father looked old and shrunken, his eyes dark-ringed in a stubbled face.

'Hallo, Mr Dawson. We were friends of Jed. We've come to say how sorry we are.'

They followed him into the tiny living-room, lit by an orange-shaded table lamp and full of people.

'Vinny, here's two friends of our Jed's come.' His voice was coaxing, like someone addressing a sick-bed, and the woman in the chair by the fire did seem in the grip of some strange malady.

'She's doped,' one of the bystanders whispered to Jenny. 'The doctor came this morning and gave her something to put her out.'

'We came to say . . .'

'Yes,' Barbara was joining in, 'we wanted you to know . . .'

'. . . how much we loved him . . . liked him. He was the

47

best friend we could ever have had. The others are coming, if they haven't been already. We all want you to know . . .'

'And if there's anything at all we can do . . .'

'Here's a cup of tea, lass.' The tea-bringer was young and huge and kindly. 'Find a chair, I'll bring some tea for your friend. Lavinia isn't saying much, but she'll be glad you came.'

'He was a nice lad,' another woman said. 'And clever. Bright as a button even as a little lad.'

'I mind on when he wrote that poem about the Queen . . .'

'And the time he tried to walk the pit pond.' They were off, then, in a welter of reminiscence. Jenny sat smiling, out of politeness at first and then out of genuine interest. She did have a bond with these strangers: all of them had been Jed's friends. She sat on, rapt, until Barbara tugged at her arm and it was time to go.

'Wasn't it *awful*?' Barbara said when they were back in the car. 'All those people in that tiny room with that great big fire. What are they all doing there?'

'They're just showing sympathy . . . solidarity. Like we were. Jed would've liked that.'

'Well I think they were just there to gawp. Honestly, they're like another breed. It would never happen in Sunderland, you know that.'

'You make it sound as though Belgate was another continent, Barbara. It's only five miles from Sunderland.'

'I don't care how far it is, they're just a different sort of people.'

'Jed wasn't different.'

'Yes, he was. He was . . .' Barbara hesitated, unwilling to criticize the dead. 'He was, well, uninhibited. You've got to admit that, Jenny. I mean he was always attracting attention. Kath's the same. Still . . . I don't mean to say anything bad about him, Jenny. I really did like him, whatever it might've seemed like when we had our spats.'

'I know,' Jenny said absently. Suddenly she was thinking ahead to their meeting in the coffee bar, to coming face to face with Keir. There were moments when she almost forgot what had happened between them. Now, as the car neared the town centre, she was suddenly afraid. What could they do, say to each other, that would be enough to reinforce their bond without letting the others know exactly what had happened?

In the event it was Keir who masterminded everything. 'Come on, Jen, sit next to me, you look frozen. Move along Kath, Barbara wants in.' It was better when she was in the booth, her shaking knees beneath the table and Keir's arm lightly around her, as though he were simply consoling a friend.

'We're talking about last night,' Kath said. 'No one can make sense of it. I went round Jed's this morning and they say the police said he was dancing about with someone – so he must've known them.'

'We've just come from Jed's, too . . . his parents are awfully crushed, aren't they? Has everyone else been to see them?'

'I'm taking Elaine this afternoon,' Alan said. 'She's off duty then. Keir and Kath are more or less in and out all the time, as they live so near. So, apart from Euan, we've all shown our faces.'

'Where were you last night?' Barbara said. Usually she showed Alan a fair amount of deference. Now, though, her tone was slightly aggressive and Jenny thought she saw Alan's chin come up.

'I might ask the same thing of you. I thought I was getting a lift home when I'd seen Euan off. In the end I had to leg it, and it's no short distance.'

'Well, I'm sorry. I collected the car, I came back, and I drove round and round for ages. I don't see what else I could've done. If you were there I'd've seen you.'

'Never mind last night,' Kath said, irritated. 'You all made a balls of the journey home – so what? We should be discussing what happened to Jed, not holding post mortems on trivia.'

'He knew a lot of people,' Keir said. 'You know Jed, the world was his friend. He could've met up with any one of a dozen people . . . twenty, thirty . . . and if he hadn't seen them for a while he'd have stayed to chat, late or not.'

'Down by the river?' Alan said.

'That's what's so odd,' Kath insisted.

'Well, said Keir, 'going over and over it won't help. Wait

for the inquest. They'll find out who it was on the river bank, because he was seen from the bridge . . .'

'He or she,' Alan interrupted.

'He or she,' Keir agreed.

'And anyway, it was an accident,' Barbara said. No one spoke and she looked from face to face. 'Well, it was, wasn't it?'

'Of course it was,' Alan said. 'Come on, we're all getting paranoid.'

They were on their way out to the street when it occurred to Jenny that she and Keir might not be the only ones with something to hide. Alan had not gone home with Barbara, therefore each of them had gone home alone. Or had they? Had one of them stayed – Barbara to play out her old animosity with Jed, Alan to pursue some secret quarrel?

It seemed impossible, seeing them now, going about their everyday activities. But she too looked as she usually did – a little sad, perhaps, but otherwise unaltered. And yet something had happened to change her completely. Had the same thing happened to one of them?

The coffee-bar doors swung to behind her and there was the blessed normality of the street. It was crazy to suspect her friends. They had been Jed's friends too and therefore were above suspicion.

Her mother was standing in the hall when Jenny got home, the telephone in her hand. 'It's Keir Lockyer . . .' Her mother's tongue stumbled over the name and then, as she handed over the receiver, she mouthed, 'Is that the Belgate boy?' and Jenny nodded, hoping the sudden flush in her cheeks didn't show.

'Jenny?'

'Yes, it's me.'

'I thought I'd ring you. We didn't get much chance earlier today . . . to talk, I mean.'

'No.' Jenny was suddenly tongue-tied, unable to string two words together. She fought for something, anything, to say.

'Where are you?'

'A call-box, round the corner from home.'

'Oh.'

'I was wondering what you were doing tonight?'

'I'm seeing Barbara.' The relief of having an excuse swept over Jenny. She wanted to be with Keir more than anything else in the world, but not yet, not till she had had time to sort herself out. Suddenly panic engulfed her: he might think she didn't want to see him, might never ask again. She could cancel Barbara – this was important, vital . . . the most vital thing in the world. She was opening her mouth to retract when he spoke.

'Tomorrow, then?'

'Yes, that would be wonderful!'

'The bus station at seven?'

'Fine.'

'Seven it is. Decide what you want to do. Anything'll suit me.'

When Jenny put down the phone she felt weak. She had got what she wanted, or it seemed as though she had. She went upstairs slowly, thinking at one and the same time of what she would wear and say and do tomorrow. She intended to open her wardrobe, get out her best and see what she could make of it, but in the end she lay down on the bed, turning her face into her arms as though afraid to face the light, thinking of the terror and the wonder of what was happening.

She had wanted Keir and she had got him. It would work out between them, she would make it work! Panic and elation alternated. She would be desirable – gauche – irresistible – boring. When she could stand it no more she sprang up and went in search of coffee.

The pub was half-empty – a clutter around the bar, a group playing darts. Kath was with Dave Gates in a corner booth from which they had a view of the rest of the pub, the Half Moon. They sat with their drinks in front of them, a half of lager and lime for her, a pint of bitter for him, and talked about school and childhood, teachers they'd liked, teachers they'd feared.

'You were always bright,' Dave said. There was no great

admiration in his voice: he was stating what he believed to be a fact. She liked that.

'I was average . . . and I was keen. Jed was the bright one.'

'Come on now, keep off that. It'll only upset you.'

'Yeah. Anyroad, you were no slouch yourself.' While Dave protested she pondered how easily you fell back into the jargon of home. 'Anyroad': she never said that at college, and not even in Sunderland when she was with the others. But here, in the local, with a boy she had known all her life, it had come to her as easily as breath.

'Want another?'

While Dave stood at the bar Kath watched him, seeing the broad shoulders, the lithe waist, the strong legs of a working collier. He had been good to her these last few days, since he had knocked on her door to tell her about Jed: 'I thought you should know, seeing he was a mate of yours.' She had cried, and he had gathered her clumsily into his arms, then offered to drive her anywhere on his motor-bike, if it would help.

Kath had always fancied Dave a bit, but they had seemed to be worlds apart in the last few years. It had not come about from any desire on her part; rather it was he who had withdrawn, as others in Belgate had done once she was labelled 'brainy'. She had gone into the sixth form; he had gone into the pit. She had made new friends at college; he had kept their old friends, and she had been shut out.

Now, sipping their drinks, they argued about music, discussing the relative merits of Dire Straits and The Police.

'Sting's so sexy,' Kath said, grinning provocatively.

'If you like them weedy,' he said equably.

'Ooh, miaow, miaow,' she said.

'Nee chance, Kath. I may not be much but I'm his match.'

'What about girl singers? Who'd you like?'

'Elaine Paige . . .' He was grinning. 'And Dolly Parton. The voice's not much but she's got compensations.'

'Oh God, men . . . you all flap at the sight of a big chest.'

'I didn't say that,' he said. 'I just said "compensations". Anyroad, what you like to see at the pictures isn't mebbe what you want at home.'

Kath pulled a face and glanced down at her chest. 'There's hope for me then?'

'I daresay someone'll take you on, if you play your cards right.'

Kath suddenly realized that she was happy, here, on her own home ground, making easy conversation about nothing in particular.

'Thanks a bunch,' she said and grinned at him above her glass.

'It sounds like a little paper,' Barbara said dubiously.

'It's a provincial paper,' Jenny replied. 'It's not *The Times* but it's not that small. Anyway, it's a start.'

'And right down there in Coatham? I mean, who will you know? What will you do with yourself? You'll be completely cut off.'

'It's 1980, Barbara. There's the train. I might even get a car.'

They were sitting in Barbara's bedroom, dolly-print walls and curtains, matching bedspread and easy chair, a white bookcase topped with the dolls and soft toys of childhood. The room throbbed to a Diana Ross album: *'Touch me in the morning, then just close the door . . .'* If only she could tell someone about Keir. But she couldn't tell Barbara – not all of it, any way.

'Put that off, Jenny,' Barbara said suddenly. 'It makes me think of Jed. I mean, it makes me feel emotional, and then I start brooding. I wonder when we'll hear from Euan?'

'Soon, I expect. If his mother's told him.'

'Why shouldn't she?'

'Oh, I don't know,' Jenny said. 'But let's face it, Ba, they sent him away to separate him from us, from the life he was leading. They're hardly likely to risk it all starting up again.'

Barbara's eyes were round. 'I don't see how they could object to us. Not you and me, or Alan. The others maybe – the Belgate people, or Elaine. Especially Elaine. But Euan never bothered with her.'

'He did! He was very fond of her.'

'Not really. I mean, he was polite and nice to her, but that was just good manners, the way he was brought up.'

'I think Euan is naturally courteous. He's got all the little airs and graces but underneath he's just thoroughly nice.'

'He's thoroughly rich. Dad was saying the Craxfords own half of Newcastle.'

So Euan's pedigree had been discussed. Jenny suppressed a smile.

'I suppose he is a good catch.'

'Too wild though.' Barbara's face was serious. 'Money's important . . . essential even. But you want someone reliable . . . well, I think you do. Euan doesn't know where he's going, he dropped out of university, he never even thinks about a job or responsibility. He's simply not reliable. And he drinks too much!'

Was Keir reliable? Jenny half-listened as Barbara chattered on about eligible men, thinking all the while of Keir and the way his eyes seemed to eat you up when he spoke to you.

'Jersey's full of them. I mean you can't live there unless you're solvent. That reminds me . . .' Barbara was reaching for her bag and taking out her flowered notebook. 'I want to book some sun-ray treatments. We don't want to arrive there looking like something that's crawled out from under a stone.' She wrote studiously for a moment and then sucked her pen. 'We'll have to get it all worked out, Jen. It'll be the holiday of a lifetime if we organize it properly. I've made a list of what I want to wear . . . I know what I want, I haven't worked out how I'm going to pay for it. Still, we've got five months.'

How could she go to Jersey and leave Keir? Panic gripped Jenny for a moment, but she could always make an excuse. Jersey was months away, the finals a mountain range to cross – and if Jed's death had taught her anything it was that you never knew what the future held, so why waste time in making plans, or fighting plans others made for you?

5

21 January 1980

They had agreed to go to the inquest together, assembling outside the grey court building, not speaking much, Alan unobtrusively squiring Elaine, Keir with Kath because they had travelled to Durham together, Jenny and Barbara clinging as once they had clung in Infants' School.

'Are you sure they let spectators in?' Barbara asked.

'For the fourth time, Barbara, yes!' Alan was terse. 'That's the whole point of an inquest, that the public *sees*.'

'All right, I only asked. No need to go down my neck.'

They filed in, surprised to find a sprinkling of people already there. Some were familiar, like Jed's parents; most of them were strangers. There were three uniformed officers, two in the front row and one walking about, obviously in charge.

'That's the coroner's officer,' Alan whispered. He was sitting on her left, taking a keen interest in what was going on. 'The other two will be witnesses.'

'You mean the police *saw* it?'

'No, they'll give evidence about finding the body and the autopsy and what we told them.'

Jenny was at once impressed with his knowledge and chilled by his detachment. It was Jed they were talking about, funny Jed who could never be 'the body'. 'Alan has a legal mind already,' she thought but in the back of her mind the word 'cold' danced about. 'Cold' which meant chilly and could even, in certain circumstances, mean cruel. The old suspicions started up in her mind until she tidied them away at the coroner's entrance. Now they would find out the truth.

One of the policemen testified first. He had been called to the Milburn Gate bridge on the night of Tuesday, 15

January at eleven-fifty-one p.m. There, a Mr E. R. Cummings had told of witnessing an incident on the river bank below the bridge. This witness would give evidence at a later stage. 'Once I was satisfied that there was someone in the river I called for assistance. At three-twenty a.m. a body was recovered from the south bank of the river a mile down from the bridge. The deceased was later identified as John Edward Dawson, a student aged twenty-two, formerly residing at 17 Forbes Terrace, Belgate. The body was taken to the District General Hospital where death was certified by Dr O. K. Dwyer.' Evidence of cause of death would be given at a later stage by Dr R. P. Gupta, a pathologist at the same hospital.

The coroner thanked the policeman, and then the witness from the bridge took the stand. He was obviously uneasy but as he began to tell his tale the coroner was at pains to put him at ease.

'Just tell us in your own words what you saw.'

'I was walking home over the bridge. I'd been out for a drink . . . just a couple of pints. It was a cold night. When I got on the bridge I could hear the water gushing underneath, and I was half-way across when I heard something else. I thought it was voices, and I stopped, but I could only hear the water then, so I went on.'

'How much traffic was there?' the coroner asked.

'Not much. It was quarter to twelve and things were slowing down. There was the odd car. I was the only pedestrian.'

'Thank you. Please go on.'

'I walked on but then I heard it again. Shouting, or laughing, I couldn't be sure which. I walked to the parapet and then I saw them.'

'Them?'

'There were two of them. One on the bank, just standing. He was half-way back and it was dark there. On the river it seemed lighter because of the water. It was running fast over the weir and the foam seemed to reflect the light. At first I thought it was just the tree. It was caught on the bank, sticking out into the river where it'd been swept down. And then I saw there was someone on it. A man I think but I couldn't be sure. He . . . or she . . . had a long

56

coat or cloak or something . . . it seemed to be caught in the branches. He was laughing. Well, it sounded like laughter. And then he started shouting and I saw that the end of the tree . . . the roots . . . was swinging with the current. He was shouting something – "Help" I suppose, I couldn't hear.'

'What was the figure on the bank doing?'

'Just standing. He just stood there and then after a bit . . . a minute I suppose, or thirty seconds . . . he turned and walked away.'

'Where did he go?'

'I don't know. I looked back at the man on the tree. He was trying to get back on to the bank but he was caught somehow. And then there was a kind of groaning noise and the tree went down river with him on it.'

'And the other man?'

'I started to shout and I looked back, but he'd gone. I saw the tree come towards the bridge. It went over the weir and went under and then bobbed up again.'

'Was the man still on the tree?'

'I don't honestly know . . . he might've been. I just started to run. There's a phone box just past the bridge. I dialled 999 and they told me to stay put till the police came, so I did.'

The pathologist gave evidence next. He had examined the body of John Edward Dawson. The body was that of a healthy young man. There was evidence of a recently eaten meal and a blood-alcohol level more than double that for driving within the limit. Death was by drowning. In his opinion the shock of submersion in the icy water would have rendered the deceased unable to save himself, whether or not he was a strong swimmer. However, there was a contusion on the right temple consistent with a blow from a hard object and fragments found in the debris of the wound proved to be from an ash tree, long submerged in fresh water.

'In my opinion the deceased was rendered unconscious when the tree went over the weir, and drowning occurred within minutes.'

'Thank God,' Alan said quietly and Jenny looked along

57

at Elaine to see if she too was relieved. But Elaine stared ahead, seeming detached from the whole proceedings.

Now another senior policeman was taking the stand. He spoke of the enquiries that had been made to discover the identity of the watcher on the bank. Unfortunately enquiries were hampered by the fact that no one could give a description of the mystery figure; it was not even known for certain whether it was a man or a woman. However, enquiries were still proceeding. He had also made extensive enquiries of the deceased's friends who had been with him on the night he died. They were prepared to testify that the deceased had been in good spirits before his death.

Jenny shivered, thinking of her own interview – the probing questions, the answers written carefully in the notebook. The awful feeling of guilt that had crept over her . . . and over the others, too, if what they had said afterwards was true.

The next witness was an elderly woman, the last person to have seen Jed before he reached the river. She clutched her handbag against her chest as she told her tale.

'I was putting out the milk bottle. I always do it last thing and I was late that night because I fell asleep in the chair and didn't wake up till nearly eleven. I tidied up and rinsed the bottle and then I opened the front door. This young man was coming up the street, right beside my door, and I got a bit of a shock, seeing him loom up.' She looked a little agitated. 'Well, you never know these days, do you? And he did look a bit odd. He had a long green coat on, a bit like a highwayman's coat, and then the muffler up round his face. He must've seen me start because he called out. "Don't worry," he said and he held his hands up . . . like this.' She raised her hands in a gesture of surrender. "Don't worry," he said, "I'm harmless." He had a nice smile so I thought it was all right. I said something about it being a cold night, and he said, "Never mind, missus, it'll soon be spring." That was it exactly. I went in and shut the door and I never thought anything about it till I read the *Gazette* that night . . . well, the next night . . . and it said about the green coat. So then I told the police.' She formed her lips in a self-satisfied way. 'I knew I ought to do that.'

'And you've since seen photographs of the deceased?'

'Yes. And the coat. It was him, all right.'

'You live in Mount Road, Mrs Gilbert?'

'Yes.'

'Mount Road leads up from the city centre?'

'Yes, it does.'

'And the deceased was walking *up* the road? In other words, he was walking *away* from the river?'

'That's right.'

'Did he seem in good spirits?'

'Yes. He seemed happy really. I'm sure of that.'

The coroner thanked her and she resumed her seat with the self-conscious air of a public benefactor.

'The verdict was 'Death by misadventure'. There was no other possible verdict, according to the coroner. The deceased had been in good spirits according to witnesses. Whatever the relationship between the deceased and the figure on the bank, there was evidence that the unknown man or woman had played no part in the tragedy. The coroner considered that Jed's weight had dislodged the tree, and he added a little homily on the evils of drink and the waste of a promising young life. He extended the court's sympathy to the deceased man's parents, and rose to go.

There was the sound of a stifled sob and then Jed's father was leading his mother out of the court, the coroner's officer solicitous behind.

'She'll be OK,' Kath said. 'She's got good neighbours and they've always been close, her and her man.'

'We can't all get in Barbara's Escort,' Alan said as they stood outside the court.

'Yes we can,' Kath said.

'There's six of us,' Barbara said, scandalized.

'Oh come on, Ba-ba,' Kath said, pushing her towards the car-park. 'We need to talk, all of us. We can't split up now.'

Keir was ordered into the front passenger seat. 'You're the biggest so stand back till we get settled.' Kath marshalled Alan in first and Jenny next, then climbed in herself. 'Right, now Elaine, you get on my knee.' And then, when Elaine had climbed gingerly into place . . . 'My God, you don't weigh more than good-sized cat.'

59

'I'm sure this is against the law,' Barbara said as she turned on the ignition.

'Shut up and drive, Barbara. If we get stopped I'll say you did it under duress.'

'It's still my responsibility.'

'Put a sock in it, Babs.' This was Keir at his most commanding. 'You're hovering about in the shadow of the prison here. Let's get out of Durham.'

They climbed out of the car on the edge of Sunderland town centre, and made for Luigi's, leaving Barbara to park the car. 'Don't start talking till I get there,' she said as the last one unwound from the back seat.

'If we do I'll make a list of what's said,' Alan suggested, straight-faced.

They ordered coffee and sat either side of the plastic table.

'So,' Alan said.

'So what was he doing down by the river?' Keir offered.

'He's walking away from the river, all set to go to his aunt's,' Jenny said, 'and then suddenly he changes his mind.'

'Or has it changed for him.' Kath's lips were pursed.

'Not Jed,' Keir was shaking his head. 'And besides, you can perhaps persuade someone to go down to a river bank against their will, but how would you persuade them out on to a rotten tree-trunk in a swollen river? No, that was pure Jed. He was larking about once too often.'

'But why didn't the person come forward?' Jenny said. 'If there's nothing sinister, why not come and give evidence?'

'Perhaps they don't know about it.' It was the first thing Elaine had said since they left Durham and they all turned to look at her. She dropped her head at their gaze and then lifted a hand to wipe her nose. 'It was just a thought.'

'It's a possibility,' Alan said soothingly. And then, 'Your coffee's getting cold – drink it up.'

Barbara arrived, rattling the car keys, pushing in opposite Jenny.

'I'll get you a coffee,' Keir said. 'Seeing that you risked arrest to get us here.'

'I should think so, too,' Barbara said but she looked

pleased. 'Now, what've you been saying? Don't leave out a word.'

'What did you think of it?' Keir said as they filed out of the cinema.

'A bit overwhelming,' Jenny said.

Keir grinned. 'It wasn't called *Apocalypse Now* for nothing. Still, it had nice touches, like the helicopters going in in time to Wagner. I liked it!'

'Too much violence,' Jenny said.

'Perhaps it wasn't the best film to see today,' Keir agreed. 'This morning was enough realism to last a year. I still can't believe Jed won't be on the bus tonight: "Where've you been, mate?" And if I'd said I'd been out with you he'd've said, "Lucky sod." He liked you, Jenny . . . just as he didn't like Barbara.'

Jenny's eyes filled. 'I liked him. What makes it worse is that he was the most alive of all of us. Do you know what I mean? He lived every single minute . . . intensely.'

'I do know what you mean. And he was the brightest of us, probably; him or Alan. But with Alan you never know. He's like an iceberg, two-thirds below the surface.'

They caught the bus to Jenny's stop, and stood to one side of the gate, hidden from the house by the hedge.

'OK?' Keir said, bending slightly to meet her eyes as he pulled her into his arms.

'Yes,' she said, her face against his shoulder.

'Kiss me then.' She lifted her face, closing her eyes and puckering her mouth. Nothing happened and after a moment she opened her eyes to find him grinning at her.

'Am I so ugly you've got to close your eyes?'

'No, silly.' Didn't everyone close their eyes when they kissed? She always had before and no one had commented. On the other hand she had never been kissed before as he was kissing her now. She felt her mouth open and then his teeth against hers, his tongue reaching, flicking. She waited for the revulsion she had felt when it had happened with others, but it never came. Instead she felt her legs grow physically weak, so that she sagged a little inside his arms.

Wild thoughts of where they could go to be safe from

61

prying eyes chased through her head. 'I want him,' she thought, 'never mind the consequences, or rights and wrongs. I want him.'

'I've got to go now, Jen, or I'll miss the last bus. See me tomorrow?' Keir had settled it. They arranged to meet at seven-thirty in the Market Square and he stood at the gate until she reached the door and inserted her key. She turned to see him smiling indulgently, rather as her father smiled at her sometimes but with an extra exciting quality.

'Sleep tight. See you tomorrow.'

'Yes. 'Bye.' She wondered if she should blow him a kiss, but decided against it, and went into the house.

She was shutting the door as the phone began to ring.

'It's OK, I'll get it.' From the living-room she could hear the drone of the television.

'Sunderland 565656.'

He sounded as though he was here, in the next room, and yet he was thousands of miles away.

'Jenny? It's me, Euan!'

'Euan? Are you all right?' Jenny didn't know why she asked that but it seemed appropriate.

'I'm all right, Jenny. Just devastated . . . and so sad. I wish I was there with you all. Can you do something for me?'

'I'll do anything, Euan. Just say.'

'I want you to get flowers for the funeral for me. Spring flowers . . .' His voice broke and there was silence for a moment. She was just about to speak when he began again.

'Spring flowers. And on the card: "*Best mates always, Euan*".'

'"*Best mates always*" – I've got that.'

'I've sent you a money order. If it's not enough, let me know. Is someone taking care of Elaine?' Again Euan's voice broke. 'Jed really loved Elaine.'

'I know, Euan. Don't worry, Alan is making sure she's all right, and I will too. She'll be OK. It's you I'm worried about, you're so far away from us.'

'I'm OK, Jen. Broad shoulders and all that. Besides, you know me, I'm too thick to take much harm.'

Jenny wondered if she ought to tell him about the inquest

but she wasn't sure how she could find the words. Instead she said, 'How was your flight?'

'Fine. Boring, actually, but I survived. Give my love to the girls . . . and tell Alan and Keir . . .' Again his voice broke and there was silence.

'Oh Euan . . .' Jenny couldn't think what to say to him. 'I'll write. I'll write to tell you about the funeral. And I'll get the flowers for you.'

Long after she had put down the phone she could see Euan in her mind's eye, his face grief-stricken beneath the red curling hair. He was always running himself down but she knew him better than he thought, and she knew that he was hurting badly from the loss of a friend.

Steven was first in bed, propped on his pillows. He liked to watch Cissie get ready for bed, had always liked to watch her. She had changed with the years but in some ways he loved the changes. There was something more vulnerable about the sag of arms and breast and belly. She had served him well over the years, as wife and lover. And above all she had given him his son.

He thought of the boy, asleep now in the next room. Another six months and he could begin his climb – had already begun it, in fact.

'Come here,' he said, as Cissie climbed into bed.

'Wait on a minute, I'm not in yet. Hold your patience.'

'I'm holding more than me patience.' They did not make love so often nowadays but that only served to make it more intense. He pulled her down against him. 'I can't wait for you, Cissie. You've got me clamming.'

'Will you lower your voice, you'll have our Keir hear you.'

'Never mind about him, he has his own life. And life's short, Cissie. We have to make the most of it.' He could sense her smiling in the dark, could feel her heart beat through the silky stuff of her nightgown. He tugged at it until she sighed, pretending surrender, and began to pull it up over her head.

'You get no better,' she said, when it was discarded and thrown to the floor. 'Keep those covers around me, it's freezing.' He could feel her, thigh to thigh, belly to belly, her breasts soft against his chest.

'Oh God, I love you, Cissie Lockyer.'

'I should hope you do after thirty years or more.'

'I love you more now.' It was true. He loved everything more now: the air, the sea, the faces of his fellow men, his son, his wife. As though he realized only too clearly that a time must come to relinquish all of them.

He felt afraid, then, and sought succour in the only way he knew how, raising himself above her, subsiding dutifully when she said, 'Not yet', reaching to touch all the familiar secret places that were his alone to explore. It would be all right. He was still a man, a force to be reckoned with.

6

23 January 1980

'Somebody has to go to the house,' Alan said. 'We can't all crowd in but at least two of us should go to the church with the family.'

'It ought to be Elaine,' Kath said, and Jenny agreed, but Elaine shook her head.

'I haven't got the right,' she said and they could see she meant it.

'OK,' Kath said, 'I'll go. I've known him longest, next to Keir.'

'Yes,' Keir said, 'we were good mates. I'll go with Kath.'

So the Belgate pair went in while the others stayed outside in Barbara's father's car.

'Do I look all right?' Barbara said. 'I borrowed this from mum.' Jenny too had searched her wardrobe for navy or grey, and wore an odd assortment of skirt and jacket. Only Elaine had black, and to spare. Her face stood out, pale and blue-shadowed, beneath hair drawn back and caught on the crown with a comb. 'She looks like Ava Gardner,' Jenny thought suddenly, remembering *The Barefoot Contessa* on the TV a week ago.

Alan was silent in the front seat, oblivious of their worries over mourning clothes. He had seen to flowers, collecting from each of them and writing the card: '*In affectionate remembrance of a wonderful friend*'. It sounded so pompous but no one could think of a suitable alternative. They had all signed the card, and somehow the clustered signatures made the message more personal.

'I've brought some Kleenex,' Barbara said and handed them out with unaccustomed generosity, even to Elaine.

When the funeral cortège formed up they watched, seeing Kath and Keir, strangely subdued, bringing up the rear, both with downbent eyes and hands clasped in front of them.

'I hope I can keep the speed down enough,' Barbara said, biting her lower lip and gripping the wheel till her knuckles stood out beneath her borrowed nappa gloves.

There was room in the pew beside Kath and Keir, so they filed in there, bowing their heads for a moment and then sitting back to look around the church, avoiding the flower-strewn coffin until it would be ignored no longer and drew their eye.

Impossible to imagine Jed confined, Jenny thought. Neatly laid out, feet coming almost to a point, arms folded at the widest part. Was he smiling in death as he had smiled in life? Was his hair tidied for the grave as it had never been for the business of living?

'Oh love that will not let me go, I rest my weary soul in thee, I give thee back the life I owe, that in thine ocean depths its flow may richer, fuller be.' Jenny sang the unfamiliar words, hearing Kath's voice too, seeing Alan merely mouthing the words, suddenly hearing Elaine's voice soaring away. 'She can really sing,' Jenny thought, and was surprised.

'Let us not mourn for John Edward,' the vicar began.

'Jed . . . his name was Jed,' Jenny thought angrily and Kath turned and met Jenny's eye to show agreement.

'His life was short in years but rich in achievement. I have seen tangible evidence of that achievement in the last few days. Certificates, trophies . . . but I have sensed much more. In the conversations of his family and friends I have formed a picture of a young man who understood the gift of life . . . a gift we are all given and which many of us fail to appreciate. But John Edward . . . Jed, if I may call him that . . .' Jenny's eyes filled and beside her she felt Barbara tense. '. . . filled every day of his short life to the brim, with comradeship, with kindness, with laughter . . . with sheer joy of living. And let us not forget that he, in his turn, was loved . . . every day of his life. From the moment of conception he was *loved*. And the good news is that there is no need to cease loving him now. He is there, in your hearts and minds. He stays there, your son, your cousin, your friend, the recipient of your affection for as long as you choose to cherish your memory of him.'

Barbara was crying now but Jenny stayed calm, unwilling to lose a word. It was true: they had Jed still, the essence of him. Nothing could take that away unless they chose to relinquish it. She looked past Alan to see if Elaine, too, was comforted and saw that she looked happier than she had for days.

'*The King of Love my shepherd is, whose goodness faileth never* . . .' And then they were filing out and back into the cars.

'OK?' Alan said to all the girls, but Elaine in particular.

'I'm sorry. I can't stop crying,' Barbara said, hiccuping softly as she eased the car into the cortège and moved slowly out through the church gates. At the cemetery they stood shivering while the words were said and the body was committed to the ground. When it was over they lingered beside the raw earth and the gaping hole while the other mourners turned towards the cars.

'You're coming back to Jed's, aren't you?' Kath whispered on her way past them with Jed's family, and the four looked at one another and then nodded.

There was no need to hurry now and they waited until the official cars moved away. Beside the raw earth the flowers were piled like a pirate's treasure, yellow and blue and pink and white.

'Euan's flowers are lovely,' Barbara commented. Jenny nodded but forbore to say she had chosen them.

They were moving away from the grave when the woman approached them. She was young and tweed-suited, her hands gloved and a small suede tam-o'-shanter set on her blond head.

'Jenny?' She was looking at Barbara.

'I'm Jenny.'

The woman held out her hand. 'I'm Sara Craxford. Euan is my son. I saw you all in church and I knew he'd want you to know I was there too, representing him.'

'How do you do. This is Barbara, Alan . . . and Elaine.'

Sara Craxford smiled and held out her hand to each of them. 'Euan has spoken about you all a great deal.'

'Is he all right?' Alan said. There was a faint hostility in his voice and Jenny was suddenly glad that Kath, forthright Kath, was not there with them to be openly critical.

'I think so, though it's early days. My brother farms in

67

Rhodesia. Euan has not yet found his forte, and hopefully he might discover an affinity with the land. If not, we'll have to think of something else.'

She didn't sound like a tyrant but there was a formidable certainty about her every word and gesture.

'It's very sad about Jed,' Sara Craxford said, turning to go. 'A very bright young man. He came home with Euan several times. I've written to his parents. Please accept my sympathy – I know you'll miss him as a friend, and Euan will too.' There was a Land Rover parked outside the gate and she walked towards it. 'I must go if I'm to be home for lunch. It was good to meet you all. I'm sorry it had to be on such a sad occasion.'

'Wasn't she classy?' Barbara said when the Land Rover had departed, and they were climbing back into the Escort.

'She was young,' Alan said.

'More like Euan's sister than his mother,' Jenny added.

They looked towards Elaine but she was looking out of the window crying quietly, and did not seem disposed to answer. Jenny put out a comforting hand to cover Elaine's icy fingers.

'Are you OK?' she whispered. In the front seat Alan was directing Barbara on to the Belgate road.

'No.' Elaine was shaking her head and the tears were coming faster.

'No? What d'you mean?'

'Don't talk now. Don't let Barbara hear,' Elaine mouthed. 'Come round to my place tonight. I'm not working.'

Jenny was meeting Keir at seven-thirty but this was too important to ignore. 'OK, I'll be there about half-past six.'

Elaine nodded and wiped her eyes with a tissue. They didn't speak again until they were getting out at Jed's front door and composing themselves for the ordeal ahead.

The house was quiet when Alan let himself in. 'Hallo, cat.' He followed it into the living-room and saw his mother sleeping in her chair. One of his shirts was on her knee, the work-box open beside her. He stood for a moment, looking down at her. She looked older now, her hair white at the temples and thinning on top. As he watched her

68

false teeth slipped down, giving an oddly horse-like appearance to her face.

He moved swiftly away, knowing she would be mortified to wake at a disadvantage and find him watching. He went into the kitchen, closing the door gently behind him, and filled the kettle. Through the window the garden was bare and deserted, not even a bird on the spindly apple tree. He brewed tea and carried it through to the living-room.

May Denton woke when he touched her arm.

'Tea.'

'Oh, lovely.' She put the shirt aside and took the cup. 'Did it go off all right?'

'Yes. The vicar spoke well.'

'And his parents?'

'What you'd expect. He was their only son . . . only child. But they were brave. They'll be OK, I think.'

'And they've got each other.'

Alan had avoided saying that in case he upset her, but now she had mentioned it herself he seized his opportunity. 'Do you miss Dad?'

May put the cup precisely into the saucer and moved it into line.

'Well, of course. But you come to terms with everything in time.'

'I can't remember him very well. I *do* remember him . . . sitting in the chair reading the paper. And if I disturbed him he'd say, "Go and ask your mother."' Alan grinned, seeing a glimmer of a smile on his mother's face. 'Quite nicely, he'd say it, but quite firmly. And you used to say, "Don't disturb daddy."'

'Well, he was . . . he needed his rest. He wasn't well for a lot of the time.' She put down her cup and picked up the shirt. 'That was very nice. Now I must get on if you're going to have everything ready for leaving.'

'Thank you,' he said.

'I've just been putting your laundry in your room. I think that's everything now, but there's still time if there's something special you need doing.' She hesitated. 'Alan . . .' He was smiling at her and she suddenly found the words.

'I've never talked much about the past. Well, I've never

seen the point. There's only now, really, isn't there? But your father would be so proud of you.'

'What was he like? I've always wondered . . .' As Alan spoke he could sense her retreat.

'Oh, very nice. A gentleman. He liked things nice.'

'How did you meet?' That had thrown her.

'The usual way. Well, usual for those days. I had a friend who worked for him . . . it was all quite simple. Anyway, to get to the point – your father wasn't a wealthy man but there was some money. A sum. It was meant for you, for your future. I've managed on the pension over the years . . . well, we have. You've always been a good boy over money, never asked the impossible. So that money's stayed, and grown a little bit. It's there for you now, when you leave Oxford and your grant stops.'

Alan felt a sudden surge of relief. It had seemed so impossible as it drew near, living in London until the Bar Final, the dinners to be taken, all the accoutrement of the bar, sombre and old-fashioned and costing a small fortune.

'That's a pleasant surprise. Is it much?'

There was pride in May's voice when she spoke. 'Three *thousand* pounds.' She emphasized the middle word as though it meant billions.

'Three thousand pounds,' Alan said, trying to sound mightily impressed. 'That should be an enormous help. I'd've managed anyway, as I hope to get one or two small jobs. But this will make all the difference.'

Suddenly he leaned to kiss her brow. It was dry and papery and smelled of the green soap that pervaded the bathroom. 'Thank you. There must have been many times you could have used that money. I appreciate it.'

It was only when he was half-way up the stairs that Alan realized he had still learned nothing about his father and their life together.

Elaine's flat was on the ninth floor. It had underfloor heating and a view of the river and very little else.

'You take the chair, I'll have the bean-bag.'

'I'll look at the view for a while, if you don't mind,' Jenny said. 'It's fantastic.' Lights twinkled everywhere over the whole of Sunderland, and as Jenny looked she began to make out landmarks. 'It must be marvellous on a clear day.'

'It's nice in the summer,' Elaine said. 'It's a bit lonely up here, though. No one talks much. I know the woman next door to say hello to, that's all.'

'Couldn't you live at home?' Jenny asked tentatively.

'No,' Elaine said, and then, feeling that was not enough, 'I had stepfather trouble. You know.'

'Oh,' Jenny said.

'Would you like a coffee?'

'Please. Milk, no sugar.'

While Elaine was in the tiny kitchen Jenny looked out, seeing the main street of Sunderland and the bridge, and beyond that the lights of the seafront. 'Elaine's like the Lady of Shallot up here', she thought, trying to remember the words of the poem.

'Tell me what's wrong,' she said bluntly, when Elaine came back with the coffee. 'I'm meeting someone in an hour and I want to help you if I can.' To her relief Elaine didn't ask who the someone was.

There was silence, and Jenny tried again. 'I know you're sad about Jed, but it's more than that, isn't it?'

Elaine nodded.

'Well, what then? You can tell me.'

'I'm pregnant,' Elaine said.

'Oh God.' Jenny tried desperately to think of something cheerful to say. 'Well, at least it's something of Jed's. And his parents will help . . . and I will . . . and Kath.'

'I don't want anyone told, Jenny. Not yet, anyway. *Please* don't tell anyone . . . especially not Barbara.'

'You know I won't.'

'I've got to be sure first. And then I've got to decide.'

Jenny was just about to say 'decide what' when she answered her own question.

'Yes,' she said. 'Well, you can do a pregnancy test to start with. They're quite easy, and you can get them at the chemist.'

'Have you done one before?' Elaine's expression was shocked.

71

'No, but I've read about them in magazines. And there'll be instructions. I'll get one for you, if you like?'

'Oh Jenny, you're such a pal.'

Tears were coursing down Elaine's cheeks now, and Jenny moved to hug her. 'Never mind, we'll sort it out somehow. Don't cry. It could be worse.'

'Not much,' Elaine said, sniffing hard. 'It's a mess, Jenny, you don't know how big a mess it is. Still . . . I've got a kit, in fact. I just need a hand to use it.'

While Elaine went to collect the pregnancy test Jenny turned back to the window.

'This is my town,' she thought, 'my birthplace.' She could see the bridge and the cranes that marked the river's edge, the river where once they had built ships for the whole world. Somewhere down there her father's office occupied a corner site and everywhere there were spires and tower-blocks rearing up from the thousands of houses. Sunderland was a big town, spreading north and south and west, eating her hill, nibbling the edges of Durham County, held in check only on the east by the mighty North Sea.

She closed her eyes for a second, thinking of Durham, hidden now from view but clear in her mind's eye, mostly green but sprinkled with towns and villages, dominated always by the getting of coal. She had grown up without realizing that she lived in the heart of a coalfield. And then she had met Kath and, through Kath, Jed and Keir, and had found in the Belgate youngsters a greater kinship than she felt for Barbara and Alan, brought up like her in a Sunderland suburb. There had been tiny differences of language, and they had been less protected and therefore more mature, but right from the start she had felt an affinity with them.

Elaine came to stand beside her. 'It's quiet now. You should see it in daylight when the sea-birds are flying – some of them are the size of eagles.'

'Seriously?' Jenny asked.

'They're always around.' Elaine was unconcerned. 'And in summer, when you've got the windows open, you can hear them a mile off, screeching and screaming.' She

72

paused and considered. 'It's funny, really, but I've never been out of earshot of sea-birds, never in my whole life.'

'Have you never been away from here?' Jenny asked.

'No. Unless you count day trips.'

'You've never had a holiday?' Jenny was scandalized.

'No. I've never been further than Middleton in Teesdale. That's in Yorkshire, isn't it? Or is it Durham? I don't know. Anyway I went there once with the school.'

She held up the oblong box, peering at the small print.

'Your hand's shaking, Elaine.'

'I know. I'm a bit on edge.'

'Let's get it over then.'

They unpacked the test and set it out on the coffee table.

'You need a sample of urine,' Jenny said, reading the leaflet. 'In here.' Elaine took the container with a grimace and made for the bathroom. 'I must do something about contraception,' Jenny vowed. It could be her doing a test in three weeks' time. She might be pregnant now, for all Keir's promises!

'Will this do?' Elaine was back.

'Yes. Put it on the table.' Jenny added the liquid as directed and put the bottle back in the box.

'What happens now?'

'We wait.'

'Oh.' Elaine subsided on to the bean-bag, clutching her coffee.

'Have you decided what you'll do if . . .'

'No.'

'It's difficult.'

'It isn't that I can't decide – I can't even *think* about it, Jenny. When I know . . . if it's true . . . then I'll decide.' She thought for a moment. 'It's what's best for the baby. If there *is* a baby. I don't want it to have a life like me, pillar to post. I've sometimes wished my mam'd believed in abortion.' She looked defiantly at Jenny. 'Many a time I've thought I'd've been better off not born.'

'Oh, Elaine . . .' Jenny said, uncertain of what to say next.

'That's what was nice about Jed. I mattered to him.'

'You'd've got married sooner or later.'

'Maybe. I didn't love Jed, Jenny. Not love . . . not like

73

that. I liked him and I loved being with him, but that was all.'

'I see.' Jenny didn't see at all but didn't dare say so. How could you have sex with someone you didn't love? Surely it wouldn't work, you would just switch off?

Elaine set her cup down on the carpet beside the bean-bag, and took a deep breath.

'It's not Jed's baby, Jenny. I know you think it is, but it's not.'

A tiny wave of revulsion ran through Jenny. So there had been another man . . . men . . . so Barbara had been right all along and Elaine was promiscuous. But the other girl's eyes were on her, waiting for a response.

'I see.' The repetition sounded hollow but she had been taken by surprise. 'Will you tell the father? If . . .'

'No, he wouldn't want to know. And he'd make a *lousy* father,' Elaine said vehemently.

'You sound as though you don't like him?'

'I don't like or dislike him. It was just . . . well, he wanted it, and I couldn't be bothered to say no.' A tear was running down Elaine's nose and she wiped it away. 'I bet that sounds awful. But it doesn't mean all that much, you just do it for them, to make them happy. It's no big deal. I'd've done it for Jed if he'd wanted it, but *he* was all for treating me with respect. Making it special.'

Inside Jenny reason reasserted itself: this was Elaine, her friend, whom Jed had loved.

'He really cared about you, Elaine.'

'I know. That's the one bright spot, that he isn't here to know about this.'

'It's none of my business, Elaine, but didn't you use contraception?'

'I've been on the pill since I was fourteen, Jenny.' It was dark in the room now but neither of them moved to put on the light. 'When Jed started coming round, and when he didn't want to sleep with me . . . well, it just didn't seem that important. So I ran out of pills and I never got down to the doctor's. I was going to but I didn't. And then there was this one night . . . two nights, really . . . with, well, someone else. No big deal, I thought, you never fall pregnant that easily. But I did.'

'Could he, the father, could he help? If you decide to go through with it?'

'Count him out, Jenny. If I decide I'm going through with it I'll do it by myself.'

If Elaine ought to be told of the trials of single parenthood, now was not the time, Jenny decided. They sat on in the dark, talking of anything and everything except the liquid in the container in front of them which might or might not contain a yellow ring. They sat on, while the building reverberated to the sounds of homecoming, doors slammed, televisions switched on, people laughed, the smells of cooking filled the air.

'I think it's time,' Jenny said at last. Elaine went to the door and flicked the switch, and then rejoined Jenny to gaze down on the unmistakable yellow-brown disc that floated on the liquid's surface.

Keir was waiting by the clock, holding out his hand.

'Where've you come from? I was watching your bus stop.'

'I called in to see Elaine. I'm worried about her.'

'She'll get over it. She and Jed were never really an item, you know. In a few weeks it would've been over. She'll forget in time.'

Jenny didn't argue. She couldn't tell Keir the truth, and without it there was no argument.

'Where do you want to go?' he said.

'Luigi's?'

'OK.' He didn't sound too keen.

'Anywhere, then.'

They moved towards the park, not speaking, her hand in his hand inside his coat pocket.

In the shelter of the trees he kissed her. 'Oh Jenny, you are sweet.' Now that she had possession of her faculties she recognized how expert were his kisses, how practised the hands that slipped the catch of her bra and crept inside her jumper. It was cold but she didn't care, not when she was here with him.

'I love you, Jenny, I want you so much. We'll be careful. Leave it to me – I've brought something with me this time.'

75

She knew what he meant. French letters. The things they'd made jokes about in school. Mysterious things kept at one end of the counter in Boots where nice folk never looked. Jenny tried to relax, to enjoy the sensations he was arousing, to arouse him in return. *'Always be a good girl, Jenny. Men don't respect you if you're not good.'* That's what her mother had said.

'Perhaps we should wait, Keir?'

'What for? I love you now. We'll be off to college in a few days, and there isn't much time.'

He was so gently insistent that she couldn't say no – not now, when she had led him to believe she would. 'You won't let anything go wrong, will you?'

'Of course I won't; leave it to me. I love you, Jenny, you don't know how much. You're such a little darling . . .' His hand was moving up, sliding inside her bra cup, freeing her breast.

'Oh Jenny, Jenny . . .' And then he was taking her hand, moving it down so she could feel the hardness she had aroused.

'Love me,' she said and let her body go limp inside his arms.

7

24 January 1980

'I like it but I can't afford it.' Kath put the sweater back on the counter and turned away.

'Anyway, we won't be doing much socializing this term,' Jenny said. 'I don't intend to lift my head, I'm just going to swot, swot, swot.'

They moved out of the store and down the Market Square, stopping at fashion windows, gazing at shoes, goggling at jewellery.

'I'm going to have an emerald when I get engaged,' Kath said when they were settled in Luigi's. 'Big as a pea, platinum set.' She wiggled her left hand and drank coffee with her right.

'I want an antique ring,' Jenny said, ruminating. 'Rubies, probably. Or pearls.'

'Pearls mean tears,' Kath said. 'Don't ask me why but that's what they say.'

'I'll have rubies then; I've shed enough tears.'

'Too much has happened this hols,' Kath said reflectively.

'It's a month since we came home. A month – and it feels like the whole world's turned upside-down.' If only she could tell Kath what she really meant.

'Jed, you mean?'

'Jed. Euan gone. Other things.' She wanted to tell Kath the truth but the words wouldn't come.

'Other things? That sounds interesting. I've got a bit of an "other thing" myself, come to that.'

'What?' Jenny said, suddenly intrigued.

'It's nothing much . . . just a guy. Dave Gates. The guy who brought me through to yours the night Jed died. I've known him since Infants' School so it's no big deal . . . but he's nice.'

'From the way you're drooling, that sounds like an understatement,' Jenny said, bursting with desire to tell about Keir, to share with Kath, who was, after all, her best friend. But at the last moment she checked herself. Not yet. Tomorrow, perhaps, before she went back. But not yet. And there was that other secret which was not her secret and therefore easier to keep.

They walked back towards the bus station.

'I don't want to go back to Northampton this term. I never do but this time it's worse. When I think of sharing with Barbara for the next six months I could fall down and foam at the mouth. She always wrings out the dishcloth and drapes it over the tap. I don't know why but that irritates me to death. And she's always spraying things and using disinfectant. She puts so much Domestos down the bog you take your life in your hands when you pee. The fumes are lethal. "What are you writing now?" I ask and she says "Putting Domestos on my list." Not caviar or Beckett's *Godot* or Omar Sharif . . . Domestos. That's our Ba-Ba.'

'Will we have the usual knees-up on Saturday?' asked Jenny. They had always had a riotous night before returning to college but this time it was different.

'It won't be the same without Jed, will it?'

'No. He was the one who pushed it.' Kath sounded abstracted and Jenny glanced at her face.

'She's thinking of Dave,' Jenny thought. 'She wants to be with him right up to the last minute . . . like I want to be with Keir.'

'Maybe we can all meet for a drink? I wouldn't like us all to go off without doing something. And Jed would've hated it.'

'Yes,' Jenny said, 'that's a good idea. We could meet about six, and split about eight. And that means we could see Elaine.'

'She'll have no one here when we're gone,' Kath said.

Words trembled on Jenny's lips but she held them back. 'We can write . . . and we'll be back at Easter,' she said instead.

*

78

'Did you see the news?' Steven said. Keir pocketed his change and raised the pint glass to his lips. Around them the pub hummed gently, too early yet for the ten o'clock crush but with the dead hour of seven-to-eight behind it.

'I saw part of it. D'you mean the British Steel jobs?'

'What else? Eleven thousand jobs – eleven *thousand*! By the end of March! It'll *do* for Wales, there's nowt else there. They stand up and they say, "Eleven thousand men out" – just like that. Nothing about dependants or the knock-on effect, shopkeepers and that sort of thing. No, it's efficiency, it's progress. God help us if that's progress – I'd rather go backwards.'

'Maybe it's just a threat to get them back to work?'

'Aye, well, there's that in it. It'd be like one of her tricks.'

Keir chuckled. 'You don't like Maggie, do you?'

'Like her? I would bloody well swing for her. Gladly!'

'She's worked her way up though, dad. Credit where it's due.'

His father's face glowed with a kind of pleasure. 'Ah, but that's what makes it worse. She's one of us and she still shits on the workers. She's a class traitor, lad, no more, no less.'

Over the rim of his glass Keir glanced around the pub, letting his father's words go in one ear and out the other. All you had to do to carry on a conversation with his father was to nod continually and look aggrieved, and he could do that with his eyes shut.

There were one or two faces he recognized . . . boys from school. One girl who looked familiar. The boys were all well-muscled now, running to fat even inside their printed T-shirts. He put a thumb inside his jeans and moved it back and forth to check. Any sign of a beer belly and he'd go back to playing squash.

'You seem thick with Botcherby's lass?' His father's eyes were speculative and Keir grinned.

'Stop fishing, dad, or you'll come up with nowt but a wellie boot. I like Kath, she's a real good friend . . . but that's all.'

'She's brainy – God knows how with that stock, mind. Stan Botcherby's always been an ounce short of a pound,

79

and he married a Kelly. But there's a good strain come up in the girl.'

'And a teacher for a wife goes down well with a selection committee?'

'I wasn't meaning that, but it's true just the same.'

'You leave choosing a wife to me, dad. I won't let you down. I'll pick a lass with good socialist connections . . . what's Jim Callaghan's daughter doing these days?'

'The less said about that lot the better,' Steven said. 'Let me get these beers in again and let the bitter stop your mouth. I wanted a clever lad, I'll admit it, but you're too bloody clever by half.'

Keir stood by smiling as Steven ordered and paid for the drinks. What would his father make of Jenny, if they met? And what would Jenny make of him? He was suddenly brought back to his surroundings by his father's urgent whisper.

'Hey up, there's Sam Corder. Mind on what I told you about keeping him sweet – his son-in-law's on the Executive.'

'Scratch there, no . . . higher up. Higher. That's it!' In the comfort of the back seat of a borrowed car Kath squirmed with delight as Dave scratched her shoulder-blade with strong, blunt fingernails.

'Better?'

'Lovely. Do the other one.'

'Yes, madam. Anything to oblige, madam. What did your last slave die of?'

'A surfeit of delights – too much nookey to you.'

'All right, big-head, I know a few long words.'

'Middlesbrough, Tottenham Forest, Newcastle United?' Kath mocked.

'Very funny. Now if you'd said "Sunderland FC" . . .'

They were silent for a moment, still lazy with the pleasures of love-making.

'Warm enough?'

'Yeah.' She snuggled down.

'Sure? 'Cos if you're not I could warm you up.'

'That's what I'm afraid of.' But she was afraid of nothing

when she was with him. He had even made her forget about Jed.

'I wish you weren't going back on Sunday,' he said suddenly.

'Me too. It feels like a life-sentence at the moment. Still, I'll survive. Did you see that on the news tonight, about that Irish woman committing suicide?'

'The one who lost her bairns a few years ago? Yeah, I saw it. Poor bugger.' It was three years since Anne Maguire's three children had been crushed to death by a car whose terrorist driver had been shot dead by the army.

'The youngest was only six months, a bairn in a pram. I suppose she couldn't get over it,' Kath said.

'Still, cutting your throat . . .'

'Would you ever do it? Not cut your throat, necessarily, but any way – tablets, anything?'

'I don't think so. I don't claim to be clever, Kath – I wouldn't be down the pit if I had any brains – but I can't see the point of doing yourself in. There's always something you can do . . . if you're alive to get on with it.'

'Doesn't anything get you down?'

'Not much. Me mam, sometimes. The deputies now and then. Not much else.'

'But what about politics and injustice?'

He was drawing her closer. 'You know your trouble, Kath Botcherby? You talk too much. Your mouth's on a twenty-four-hour shift.'

'I like your cheek. You'll be glad to be rid of me, then.'

'No, I won't. I don't know what I'll do without you.'

'It's only for a while.' Outside the car street lamps misted suddenly as her eyes filled.

'Six months? It's a bloody lifetime.'

'Well, what else do you suggest? It's that or starve.'

'Not necessarily.'

'What does "not necessarily" mean?'

'We could get married?'

'What!' Kath was suddenly bolt upright. 'What did you say?'

'I said, "We could get married" . . . wed . . . spliced . . . tied up . . . or down, whichever you like. I can afford a wife, even a bloody spendthrift like you. I make good money: power-loaders do. Besides which, there's me football money. We'd be OK.'

'But we've only known each other five minutes.'

'Twenty years, you mean.'

'You can't count that. We hardly exchanged a word.'

'Granted. But I watched you. You didn't move a muscle, but I knew. All them back lanes with all them fellers . . . "Let her," I thought. She's got to taste rock-salmon before she can appreciate caviar.'

'Now who's a big-head?'

'I'm not boasting, Kath. Facts are facts. Have you ever had as good a man as me?' He was seeking her lips again, raising her head on one huge forearm to settle it gently against the upholstery.

'No,' she said, as her arms went up around his neck. 'No, I don't think I ever have, come to that.'

8

26 January 1980

'You sure that's everything?' Jenny's mother said, eyeing the piled-high bed.

'I hope so,' Jenny answered, wondering how it was all to be accommodated in two bags and her haversack.

'I do miss you when you go back,' her mother said.

'Me too.' It was true, she did miss them, but the pain of goodbye was leavened by the excitement of getting back to friends and tutors, and even the work itself.

'I suppose in a way it's what happened to your friend Jed. It made me think of what it would be like to lose you. And daddy felt that way too, because he told me so last night. You're so precious to us, Jenny . . . that's the trouble with having an only child, you tend to cling.'

When her mother had been reassured and kissed and sent on her way to make lunch, Jenny pondered her last remarks. It was difficult not to smother an only child, but just as difficult to be one and know you had the full weight of parental hopes on your shoulders.

What would happen to her? To the others? Some lines from an Auden poem came into her head, only half-remembered.

The stars burn on overhead,
Unconscious of final ends,
As I walk home to bed,
Asking what judgement waits
My person, all my friends
And these United States.

Jenny wasn't worried about America but the fate of herself and her friends was very important indeed.

83

She sat down at the dressing-table and regarded herself in the mirror. If she and Keir made it, if they married, they could act as a focus, a haven for all the others to turn to over the years. In ten years' time they could all meet, as they had vowed to do, not on the hill but in a comfortable home with books and deep chairs and a lilac tree outside the window. She had added a piano and a dog when she heard a real dog sniffling at her bedroom door and realized this was the last time for a while that she could go for a walk and set Sooty free to run the hill.

When the hook was baited and cast, singing, out into the waves, Dave ground the base of the rod into the wet sand and retreated to where Kath had made them a resting place. She poured him coffee from a flask and handed him a half stotty-cake, warm from her mother's oven and dripping with best butter.

'Can you cook like your mother?'

'No, but I've got other talents.' Kath knew he was going to mention marriage again and it scared her so that she turned away to look at the sea.

At intervals along the shoreline other anglers were hunched into anoraks or shiny yellow parkas. All of them watched the sea, foaming today in an east wind. As Kath watched, one seized his rod as it began to twitch convulsively, and then the reel was swinging in and out as he played his catch.

'It's a big'un,' Dave said enviously, but when the fish flew through the air on the end of the line it looked pathetic to Kath's eye, only fit to be unhooked and kissed better and thrown back into the waves.

'Why do you do this?'

Dave shrugged. "Cos I like it, I suppose. I've never tried to work it out, I'm not like you, analyzing everything. I just know there's things I like and things I don't like. I like fishing, so I do it.' He looked at her quizzically. 'Is that too simple for you?'

Kath would have liked to fault his reasoning but she couldn't. 'I suppose so. It's not always that simple.'

'Not if you complicate it, Kath. People complicate things

for themselves. They want this and that 'cos the bloke next door's got it or the wife's sister couldn't live without it. Not 'cos *they* want it but because they've missed the bus if they don't get it.'

'Don't you want anything?'

'Right now I want another piece of your mam's stotty and then I want a bloody great cod to jump out and land at me feet.' She handed him the bread.

'And there's one other thing I want.'

'What?' She knew the answer but it was easier to feign ignorance.

'You know.'

'I don't.'

'Stop complicating things.'

'If you mean me, I think you've already got me.'

'Not properly.'

Kath stood for a moment, looking at the sea, grey and implacable. 'I've thought about it, Dave, ever since the other night. I couldn't give it all up, not now. I've worked too hard and, besides, me mam's set her heart on me seeing a bit of the world. I want to be something, Dave: I want a qualification, something that'll take me wherever I want to go.'

'We could see the world. Good holidays, get a caravan maybe and tour about?'

'It sounds all right but it wouldn't work out like that. We'd have a mortgage and kids and HP. I'd wind up washing for a houseful of miners, just like me mam.'

She saw he was hurt and she wanted to make amends but she couldn't deny what she'd said. She looked along the coastline to Sunderland. Jenny and Barbara would be sitting down to lunch soon in their neat little semis, packing in their nice poly-cotton bedrooms. When she got home she would have to rescue her laundry from a basket piled high with her brothers' washing, struggle to retrieve her toilet things from a bathroom that looked like a tagareen shop, and hunt down her holdall because their Ted had borrowed it for his football gear. She felt angry and guilty at one and the same time, enough to make her eyes mist so that she needed to brush them with her sleeve.

She turned away so that Dave would not notice and saw Steven Lockyer picking his way along the high-water mark.

'There's Keir's dad,' she said and Dave turned.

'I've worked with him for six years, Kath. Good worker – Compeñ. Secretary, an' all. He saw me right when I broke me foot.'

'I knew he was a big union man from Keir.'

'Oh, Keir's his blue-eyed boy, all right. Going to be a great man, according to his father.'

'He probably will be. He always was clever. Was he your year in the juniors?'

'Aye. My year, my class. Always top. We used to say to each other, when we had a test, "Why bother . . . it'll be Keir Lockyer." And it always was.'

'No wonder if the rest of you didn't bother.' Before Dave could answer the line began to whine and the spool to spin like a mad thing. He crammed the remains of the stotty into his mouth and leaped for the rod.

'This is it, Kath – an eight-pounder. You can take it home for your mam. Keep her sweet.' He turned his head to smile at her while his hands grasped the rod and she saw his feet brace themselves against the ridged sand.

'I love him,' she thought. 'God help me, I believe I really do.'

Keir and Alan had met half-way between Belgate and Sunderland, in a small pub where dominoes clacked and the barmaid knew her clients by the name of their drink. Now they stood at the bar, each with one hand loosely curved round a pint glass, the other resting half-in, half-out of his hip pocket.

'It's an expensive business, the Bar. Better you than me.' Keir lifted his glass as he spoke and let his upper lip caress the foaming head of the beer. In the background a juke box started up . . . Tom Jones singing 'Delilah' as though his life depended on it.

'I try not to think about it,' Alan said. 'If you did, if you thought of the impossibility of getting decent chambers, the struggle to make a name, the cost of all the trappings, you'd pack it in in the first week. I just keep my head down and inch forward.'

'You'll be snapped up. A brilliant first, a dazzling reputation in debate . . . you'll streak through.'

'Oh yes, Lord Chief Justice by Christmas. Still, I have worked – and so have you, for that matter.'

'For slightly different reasons, my son.'

'Such as?'

'Because if I didn't come up to expectations that old bastard who fathered me is quite capable of sending me back to do it again. And I have had enough!' Keir attacked the beer again, this time half-draining the glass.

'Is he still plugging the union job?'

'Oh yes. This man who protests about the public-school Mafia is only too willing to secure a favour for his son. According to him I will get a well-paid job as a union lackey doing PR for the Guild of Saggermakers Bottom Knockers, or, if I get a good degree, I will become an agent somewhere where they weigh the Labour votes instead of counting them. All I have to do, so the theory goes, is keep my nose clean and the member's name before the public. In no time at all I will "get a good name". This will be my ticket to the Palace of Westminster. Once there, I will pass laws to prevent anyone benefiting from privilege – unless, like me, they happen to be a son of one of the faithful, in which case they can pull strings till Keir Hardie's balls fall off.'

'No need to get bitter about it.'

'*Au contraire*, old son, every need to get fucking bitter. It offends, not my conscience, which is, as you know, entirely moribund. No, it offends my ego that my father, my fucking father considers me so useless that without his machinations I will do sweet bugger-all with my life.'

'OK, so if he stayed out of it what would you do?'

Keir looked around the pub. 'I dunno. Beachcomb. Live off rich old ladies. Anything *un*-worthy.'

'If you're not careful you're going to let your father run your life. Oh, you may not do what he wants but you'll do whatever you do to spite him, which means that he's still responsible for where you wind up.'

'You're a plausible sod,' Keir said, grinning. 'If I'm ever

in trouble I hope to God you're defending, not prosecuting.'

'That's a deal. One more before we split? I mustn't be late today, not for the ceremonial last lunch. I don't know about your folks but it's a ritual with my mother.'

'I know what you mean,' Keir said. 'We'll leave it then. We're all meeting tonight, after all.'

They met for a farewell drink, five of them. 'Elaine's working tonight,' Alan said. 'I said we'd call in to the Lion and say goodbye to her.' They drank, curiously silent, exchanging information about trains and the dates of half-term and renewal of railcards. They tried to be cheerful, even wild, but somehow they had lost the knack.

'It's not the same,' Jenny thought and saw in Kath's eyes a reflection of her own mood. 'It isn't that I don't care about them,' she thought, 'it's just that Keir fills my thoughts now. There's no room for them any more.' She was suddenly filled with guilt at the fickleness of her own affections. She had always known love would come, but she had never bargained for it changing everything. She had thought it would add to life; instead it obliterated it. 'There is only Keir,' she thought. 'Everything else is second best.' It was good that he had to go back to Oxford; if he had not she might not have found the courage to tear herself away from Sunderland.

'Are we getting away from here?' Keir whispered, when Alan and Barbara had gone to the bar for another round and Kath had subsided into a reverie. He put out a hand and touched her forearm, pushing aside her sleeve. Jenny felt a sudden unbearable tingle and then her heart was pounding uncomfortably in her ears.

'We can't just walk out,' she whispered and, as if she had heard, Kath sat upright.

'I hate to break up the party but I'm catching the half-past eight bus. What about you, Keir?'

'I've had enough of this. It's not so much a ceilidh, more a wake,' Keir said.

'We can't just walk out,' Jenny said again but already she was reaching for scarf and shoulder-bag.

'Leave it to me,' Kath whispered, and then, when Alan and Barbara returned . . . 'Drink up, folks. It's our last night and I'm not packed. So if you want me on the twelve-fifteen train for Northampton tomorrow, Babsy, you'll have to let me go.'

'Me too,' Jenny said quickly. 'I haven't even made a start . . . sorry, and all that.'

'We've got to say goodbye to Elaine first,' Alan said sternly and only Barbara demurred.

'I don't see why we should keep up with her now,' she said as they trudged towards High Street and the pub where Elaine worked in the evenings. 'She was Jed's friend, not ours. We can't keep on pretending things are the same.'

'She's my friend,' Jenny said firmly. 'I've always liked her.'

'You would,' Barbara said. 'You'd find something good to say about Mao Tse-tung.'

'Which is more than we can say for you,' Kath snapped from behind. 'In your book, Mother Teresa's a Calcutta kerb-crawler.'

'Funny, funny,' Barbara said but she was discomforted all the same.

They clustered in a booth in the Lion and Elaine brought their drinks on a tray. 'I've got five minutes,' she said lighting a cigarette with a shaky hand. 'And this round's on me . . . to wish you luck.'

They drank and talked, but the conversation was stilted, an effort. 'Well,' Elaine said at last, 'duty calls . . . and if it doesn't the boss will. Keep in touch.' Her eyes glittered and she looked away. Jenny stood up.

'I'm going to the bog. Won't be long.' She put a hand on Elaine's arm and moved with her toward the back of the pub.

'This is my number in Liverpool, Elaine. It's halls of residence at the university, so anyone might answer, but they'll fetch me if you ask.'

'Ta. When will you be home again?'

'In about four weeks, I think.' It would depend on Keir's plans . . . she'd time her visit to coincide with his. But she couldn't tell Elaine that, in spite of her disappointed face.

'You're going to need someone, Elaine. Someone to talk to. Why won't you tell the baby's father? He has a right to know . . . and he might help you.'

Elaine shook her head again, this time with finality. 'No, he wouldn't want to know.' She was looking back towards the booth. 'Besides, he's got a lot on his plate in the next few months.'

It wasn't . . . it couldn't be . . . 'Is it Alan?' Jenny said and knew, even as she spoke, that it was not.

'It's Keir, if you must know. And he's not a one for facing up to things, is he?'

All Jenny could think of as they sat finishing their drinks was that Barbara would protect her when it was time to go home. For the first time she gave thanks for fussy Barbara and her car, her shield and excuse for not being alone with Keir. But it wasn't as easy as that.

'I've fixed it with Alan,' Keir whispered as he drained his glass. 'He's asking Barbara to stop on for another drink. I used the old pals' act and he gave in.'

And so it was. As Kath and Keir stood up to go, Alan leaned forward and spoke to Barbara, who turned, suddenly self-important, and said, 'You will be OK, won't you, Jen? I want to stay on for a bit. I'm already packed.'

To argue would be to cause interest, and Jenny could hardly muscle in on what Barbara was thinking of as an assignation.

'I'll see both the girls on to a bus,' Keir said as they said their goodbyes and he trundled Jenny and Kath out into the street.

'Well, I'm off,' Kath said, looking at her watch. 'Phone me tomorrow night, Jen. Only another six months, though, and it's end of sentence.' Suddenly she held out her arms. 'Give me a hug.' Her eyes filled and Jenny knew they would both cry before long. Around them the street lamps glittered, each with a pale misty aura.

'Be good, Kath. We'll be together soon . . . and I *will* phone. I might even get over to see you – Northampton's not that far from Liverpool.' She wanted to say: 'Don't go, don't leave me with Keir . . . not now.' But she must keep quiet, at least for the time being.

And then Kath was gone and Jenny was alone with Keir and numb misery was spreading upwards until she felt herself incapable of speech and could only walk silently onward at his side.

Suddenly he was taking her arm, pulling her into a shop doorway. 'Come here, Jenny. Let me warm that cold little face.' But she couldn't, wouldn't kiss him. It wasn't true, what Elaine had said – somehow, although she didn't know how, it would all turn out all right – but still, she couldn't behave as though things were unchanged from the last time they had been together.

'I can't stay, Keir. Not now. I really have got to pack. I'm sorry.'

She felt him withdraw, sensed in the darkness that his brows had come down in a frown.

'Just as you like. It's a waste of a night, though. Our last night for ages.'

'Go to Elaine,' Jenny suddenly wanted to say savagely. 'Go to Elaine and see what you've done.' But all she said was, 'I'm sorry,' again and again in different variations until she was safe on the platform of her bus and he was striding away to find his own.

9

29 February 1980

Jenny sighed and shifted in her seat. Only one more stop, and then Durham. She would not have chosen to come home this weekend, with her finals beginning to loom, but she was worried about Elaine, alone and pregnant in Sunderland. She tried not to think too much about anything except Elaine and her predicament, but painful memories forced their way in.

Keir's face in the park; the sound of wind and single raindrops on the leaves: *'Oh Jenny, darling Jenny,'*; and then Elaine . . . *'It isn't Jed's baby, Jenny. It's Keir's.'* Keir's baby alive in Elaine's womb, the death of all her dreams. Keir's baby, the reason she was coming home now for the weekend.

She had worked hard since January, burying herself in books, trying to think of nothing but the future. She would get a good degree, she would, she would! She would go to Coatham and learn the art of journalism; go to Fleet Street, get a byline; be somebody. She thought of Jean Rook, dripping mink and bracelets. If she was famous perhaps it wouldn't hurt any more. The intercom crackled to announce Darlington and she began to gather her bags together ready for alighting at Durham.

She was expecting her father to meet her but it was Kath who was waiting.

'Hi, Jen. No, don't panic, your dad's fine.' She took Jenny's bag in one hand, her arm in the other.

'I came home from college on an impulse . . . and found Dave had had a rush of blood to the head and traded his bike for a car. Delusions of grandeur, no less. So we went to Sunderland, to see if you were back, and when your dad said he was picking you up at Durham I said we'd do it for him.'

They had reached the station entrance and Jenny felt Kath's grip on her arm tighten. 'Come and meet Dave.'

He was tall and fair, already unfolding from the car to stow her bag in the boot.

'Hello, Dave. It's nice to meet you at last,' Jenny said.

'Pleased to meet you,' he countered and bobbed his head.

'It's a lovely car,' Jenny said and saw his eyes shine though the mouth remained impassive.

'We'll get in the back and gas,' Kath said. 'Dave can drive.'

'Yes, boss,' he said drily but didn't seem to mind being ordered around.

'I had a letter from Euan,' Kath said when they were under way. 'He says he'll be home in the autumn.'

'We'll have to make sure we see him, then, wherever we are. Still, we'll be earning by that time so it should be easier.'

'True,' Kath said. 'Gloriously true. Now, tell me about Elaine.' They had shared the news on the telephone but it wasn't the same as being face to face.

'She's going ahead with it.'

'I thought she might. Still, Jed's folks'll help.'

'She isn't going to tell them about the baby, Kath.' Jenny had promised to keep the secret about Keir, but it was hard not to tell Kath.

'Not tell them!' Kath turned in the seat to look at Jenny. They were passing over the bridge, Durham castle and cathedral floodlit above, but all Jenny could see was Kath's outraged face. 'She's got to tell them, Jen! I mean, it's their grandchild – she can't keep a thing like that from them, not when they've lost their son. It's the only bit of Jed they have left.'

Jenny knew Kath – knew she was capable of telling Jed's parents herself if she thought it the right thing to do. She would have to know enough of the truth to stop her from making matters worse.

'It's not Jed's baby, Kath.' Was it imagination or did even

the silent driver suddenly go still. 'They didn't . . . he
wasn't . . . well, it wasn't Jed.'

'Then *who*? I was sure Jed was gone on her, really gone.'

'He was. And she thought he was the nicest, the best
person she'd ever known. But this other thing was different, it didn't mean that much to her . . .'

'It's just wrecked her life,' Kath said slowly. 'What a
mess.'

'Don't say anything to anyone,' Jenny urged.

'You know I won't,' Kath replied. 'Do we know who the
guy is?'

'No,' Jenny said, 'it was someone we never knew.'

'Keir's home this weekend, too . . . and Alan,' Kath said
as Jenny got out at her gate. 'We're all meeting tomorrow
in Luigi's, twelve o'clock. You don't mind, Dave, do you?'

Dave grinned. 'If I do, will it stop you?'

Kath was moving round to the passenger seat again. 'Of
course. One word from you and I do as I'm told.'

'As you please, you mean,' he said, grinning at Jenny.

She watched them drive off in the car with its ancient
registration. 'They're happy,' she thought and felt suddenly envious.

Steven had seen the dog the day before, nosing into a bin
put out for the refuse cart. Today it lay on its side on the
waste ground, seemingly uncaring of passers-by.

'It's been out there all night,' a woman said, hurrying
on, purse clutched in her hand. 'I'm just off to get some
self-raising, and then I'm ringing the pollis.'

Steven paused, swishing a long twig against his thigh.
He had been glad to get out of the house, what with Cissie
keeping on about their Keir looking thin and going through
his bag for dirty washing like a terrier after a rat.

'Border Collie cross by the look of it,' he suggested. 'Just
about pegging out.'

'They don't last long this time of year,' the woman said.
'No grub and frost, that sees them off.'

'It looks like an old dog,' Steven said uneasily.

'God help it when the pollis get it. There's more kindness
in Belsen than you get in there.'

Steven looked down again. The eyes were glazed and sunk in the head, the coat matted and dirty and unmistakably grey around the jowl.

'I know whose it is,' he said, suddenly. 'They're looking high and low for it. It'd gone clean out of my mind and then I thought on.' He bent and put his hands under the dog's body.

'You want to watch it . . . they can turn nasty when they're hungry.' Her coat was flapping open over her flowered pinny and she pulled it to.

'He hasn't got it in him to bite,' Steven said.

The dog sagged in his arms like an old rug and he could see that its feet were bleeding and pink where the pads had shredded.

It grew heavy as he neared the house . . . or he was going soft. He hitched it higher in his arms, noticing the faint sickly smell emanating from its nose and mouth. It was not far off death, but at least it could die in peace. He had an empty cree and Cissie would find a bit of old blanket.

'You can't bring it in the house,' Cissie said when she saw it.

'I never intended bringing it in anywhere,' he said. He manoeuvred the dog on to the floor of the cree and patted its flank. 'Not much longer now, bonny lad.'

'Is it going to die?'

'We're all going to die,' he said stoically.

'Cheerful!' she said and turned into the house. He found an old mat in the shed and was about to cover the dog with it when Cissie came behind him.

'Get out of the way.' She had a blanket on her arm and a plastic carton in her hand and he wasn't moving quick enough. 'Mind out.'

She laid the blanket over the dog and then stroked its matted head. 'Have a drop milk, pet. Nice warm milk.'

'It's too far gone,' Steven said, but the dog had put out a tentative tongue, not for the milk but to lick the stroking hand.

'Come on then,' Cissie said, dipping first and second fingers in the milk and putting them to the animal's mouth. At first the milk trembled on the dog's lips but as she

95

dipped and anointed again and again it began to lick the milk into its mouth. Cissie turned.

'Don't just stand there, Steven. Go and put that rug down by the fire and then give me a hand.'

Keir appeared at the back door. 'What've you got there?' He inspected the dog and shook his head. 'Some folk should be shot.'

They carried the dog in to the fire, its body still limp and unprotesting. Cissie had fetched a battered spoon and began to ladle the milk between its lips.

'Watch out you don't choke it,' Steven said and she withered him with a look.

'It's not the dog I'll choke.'

'I'm off out,' Keir said. 'I'll see if I can pick up a few more strays on the way home.'

'Aye, laugh. I just didn't want it to die in the open,' Steven said.

'It's not going to die,' Cissie said firmly. Her fat white arms moved over the dog, patting, comforting, feeding, nudging it back into life. 'Take your coat,' she called to Keir as he went through the door but her eyes were on the dog.

'It'll have to go down the pollis when it comes round, mind,' she said. 'I've put up with your birds and your politics. I'm not taking anything else on.'

Steven sat down in his chair and picked up the paper, watching his wife as he had watched her twenty years ago fussing over their bairn. He'd certainly done one thing right, picking Cissie Hunt all those years ago. He'd been a daft young bugger, come to think of it, a hot-head, but he'd got that right and no mistake.

'I'll see you later, then,' Kath said. They had pulled up at her door and the engine was still throbbing quietly. Outside the afternoon light was fading but neither moved until Dave reached out and switched off the engine.

'I've got to go in and show me face, Dave, if we're going out tonight. I can't come home from college and never see me own folks.'

'I know.'

'Well, let me go.'

'I'm not stopping you.'

She leaned towards him. 'You are. You're mesmerizing me.'

'Talk English.'

'You know what I mean,' she said but after she had kissed him and was walking up the yard she wondered. *Did* he know the meaning of the verb 'to mesmerize'? He would understand what she meant but the word itself might not be in his vocabulary.

'He's not thick,' she thought, 'he can match me in most things. He just doesn't choose to use all his brain.' She felt a little quiver of disapproval until a voice in her head reminded her that he was perhaps the happiest person she had ever met, so who was the clever one in the end?

'You're back, then,' her mother said. She had flour on her forehead and a batch of proving bread stood in the hearth.

'Yes. We met Jenny at the station. She's fine. We're getting together tomorrow, all of us, Keir and everybody.' She poured herself a cup of tea from the pot stewing on the hob and sat down opposite her mother. 'One of the girls is pregnant.'

'Not Barbara?' Her mother's eyes had saucered.

'No, not tin-knickers.'

Her mother tut-tutted but her eyes twinkled. 'Not on her list, is it?' she said, poker-faced.

'Definitely not! It's Elaine . . . you don't know her, but she was Jed's girl-friend, more or less.'

Her mother pursed her lips. 'Poor lass.'

'It's not his, mam. I thought straight off it would be Jed's, but Jenny says not.'

'She's likely gone wrong with the shock,' Mrs Botcherby said. 'I've seen that before. Good girls, couldn't stand grief, looked for consolation and got landed with a bairn. You do your best for her, our Kath. And let me know if she's short of anything. I'm not Rockefeller but there's four wages in this house. We can always spare a few bob.'

'You're a softie,' Kath said, affectionately. She put her arms round her mother and hugged.

'I'm not,' her mother said defensively. 'I just don't like

97

to see a girl in trouble. Bairns are hard work when there's two of you, let alone one.'

'Were we an uphill struggle?' Kath asked teasingly, expecting to be reassured.

'Yes,' her mother said shortly. 'I wouldn't part with one of you but if I said it'd been easy me tongue'd fall out.'

'Never mind.' Kath was taken aback but she wanted to console. 'We'll all be a credit to you one day.'

'*You* will,' her mother said. 'I love my lads, but you're the one I have hopes for. Now get out of my way and let's get on.'

Behind Jenny the dog ran hither and thither, nose to the ground, delighting in the smells of the hill. By night this place would be alive with foxes, moles, rabbits and stoats, all taking advantage of the darkness. Now it was heaven for one small black dog with a curious nose. The grass was short and cropped by winter winds with here and there a clump of gorse or a nettle, spread like a flower, its centre jewelled with frost. Vikings had trod this hill, and marauding Scots; a thousand generations had climbed its sides to stand proud above the town.

Jenny looked down on the darkening vista, seeing the buildings blur and fade, the reverse of the way they had become clear on New Year's morning. Tears pricked her eyes, and she knew it was time to go home and get ready to go to Elaine's flat.

'What time will you be back?' her mother asked as she was leaving.

'I won't be late – we're not going out, we're just going to have coffee and talk. I want an early night.'

'When are you seeing Kathleen and Barbara?'

'Tomorrow, in town. We're meeting at lunch-time, but Elaine's working then. Besides, I think she needs to talk to me without everyone listening in.'

Her mother frowned. 'Is Barbara still being unpleasant?'

'She's stopped being outraged, if that's what you mean. But she's made it pretty clear she doesn't want to be involved. Elaine has brought it on herself, according to Barbara.'

'That's partially true, darling.'

'I know, but there are . . . well, extenuating circumstances.' She couldn't tell her mother the truth, not all of it. She couldn't tell that to anyone.

'Of course, and even if there weren't I wouldn't want you to sit in judgement. But you can't live other people's lives for them, Jenny. It never works.'

She thought of her mother's words as she went up in the lift to Elaine's flat. Was that what she was trying to do? She hadn't felt that she had any option but to be involved. All these weeks Elaine had behaved as though speaking to her on the phone was a lifeline: *There's no one else, Jenny. I'm sorry to bother you but there's no one else.*

So Jenny had listened and sympathized and exhorted, and had felt that she was doing it as much for Jed as for Elaine. But all the time she was remembering the park, and what she and Keir had done, and regret was upon her, sharp and painful.

Virginity did matter, after all. If only you could have it back, not for daft reasons like making a present of it to your husband: that was for Mills and Boon. You needed it for yourself – or, if it was gone, you needed to know that the loss of it had made sense.

Elaine answered the door almost before Jenny knocked. 'I was waiting for you,' she said and gestured to the chair. 'I'll take the bean-bag.'

'Let me,' Jenny said.

'Not yet, Jen. Soon enough I won't be able to get down to the floor so I might as well do it while I can.'

'Have you told your mother yet?'

'No. I went home the other day, all fired up to tell her, but she was in trouble as usual . . .'

'Trouble?'

'Debt, she's always up to the eyes in debt – clubs, catalogues, the tally man. He only gives her housekeeping when he thinks he will, and he thumps her if she complains. So I listened to all her troubles and decided I'd keep mine to meself.'

'You'll have to tell her some time.'

'Not necessarily. Not if I go down south.'

'Down south? London?'

'I've got a cousin down there, and a brother in Reading. Not that I'd be moving in, but I could always have a bed for a night or two while I look for a place.'

'But how will you manage?'

'Oh, I always manage,' Elaine said, at once perky and sad. 'Anyway, I've got to leave Sunderland.' She saw Jenny's brows wrinkle. 'I can't stop here, Jenny, not without everyone asking questions. I expect everyone thinks it's Jed's bairn and I'm not contradicting them. But what if his family hear about it? I know it's not likely, but you can't be sure, if I'm living here on top of them. And then the bairn might not resemble my side when it starts growing: it might be fair, for a start. Keir's a good-looking lad . . . different. You know what I mean. He's got a real good-looking face and the way his hair springs up . . . it's a bit unusual. I can't run the risk of someone seeing a resemblance, especially not *him*.'

'Don't you ever want him to know?'

'No.' For once Elaine's tone was firm. 'No, I never want him to know.'

'What do you want?'

Elaine thought for a moment. 'I've never really known what I wanted . . . not until it was too late, anyway. When I was a kid I used to think all I wanted was one day without a row. Me mam and dad went at it hammer and tongs . . .'

'Your real dad?'

'Yeah. He went when I was eleven, and then I thought all I wanted was to be old enough to work. We were always skint. There are five of us, including me.'

'Are you the oldest?'

'No, I'm number three. Two off the top, two off the bottom. Anyway, me mam married again when I was thirteen.' She laughed. 'I was such a little idiot, I thought it was a big romance. I got a new dress for the wedding. I was all for it, more fool me.'

'It didn't work?'

'You can say that again. He used me mam for a punch bag and me for a little bit of the other. He tried our Pauline first, but she stuck up for herself. I put up with it until I was fifteen and then I moved out. That was four years ago. I've survived that long, so I can survive a bit longer.'

'But there will be two of you this time.'

Elaine shrugged. 'I'll cope. The only thing I worry about is whether it's fair to let everyone blame it on Jed. He was so nice with me, a proper Lancelot . . .'

'Lancelot was a bit of a lad. It was Galahad who protected women.'

'Galahad, then. But do you think it's fair?'

'I don't think Jed would mind, if that's what you mean. In fact, I think he'd've liked it.'

'Who stood and watched him die, Jenny?' Elaine said suddenly.

'I don't know. I don't think we'll ever know, so it's best not to think about it. Did you get a letter from Euan?'

It was there on the mantelpiece, with its exotic stamp and air-mail franking.

'Will you tell him about the baby when you write?'

'I suppose so. But I don't want him to know about Keir: that's between me and you.'

Going down in the lift Jenny pondered the ironies of life. Of all the people to be in Elaine's confidence, why her? Why the girl who had yearned for Keir, had given him her virginity, who even now would take him back in a flash if she didn't keep a hold on her emotions? She tried not to feel sorry for herself. Elaine's lot was so much worse. The lift doors clanged open and she walked out into the starlit night.

The sooner her finals were over the better. Then, with a bit of luck, she could go to Coatham and start a new and satisfying life. She turned and looked back at the tower block, trying to identify Elaine's window among a hundred lighted rectangles, remembering the lovely face, sad now but still defiant. If Elaine could find courage, so could she.

10

2 April 1980

They were all there again, in the coffee bar, Alan and Keir, Jenny, Kath, Barbara and Elaine. Spring sunshine filled the room and the leatherette seats were new and bright scarlet, but the old easiness was gone and after the first coffee they moved to the bar of the Saracen.

'Lagers all round?' Keir asked but Elaine shook her head.

'Bitter lemon please.' She was noticeably pregnant now, not just in the swelling of her abdomen and breasts but in the roundness of her cheeks. 'It suits her,' Jenny thought and then noticed the nails, still bitten to the quick.

'Have you packed in the job at the café?' Kath asked.

'No, I've got to go on till I'm twenty-nine weeks, to get me Maternity Allowance. He's not very pleased, the boss . . . still, I let it go in one ear and out the other. I've packed in the pub. It's a pity 'cos I liked working there but . . .' She shrugged. Across the table Barbara sought Jenny's eye and her look spoke volumes.

'Oh Barbara,' Jenny thought, 'sometimes I could choke you.' Last night Barbara had held up her newly painted nails, shaking them dry, and delivered her verdict on Elaine.

'I've no sympathy, Jenny, she brought it on herself. I told you what she was and you didn't listen. Now she's going to expect the rest of us to foot the bill. She's cheap, Jenny, and I'm surprised you can't see it.'

Jenny had walked home pondering Barbara's words. Was Elaine cheap because she had had sex with someone, or because she was pregnant and had been found out? If it was the former, then she herself was cheap. If her own involvement with Keir had led to a pregnancy, would she have been even cheaper? One thing she knew now, the old myth about sex – that having once tasted it you couldn't do without it – was a load of balls. She had never felt less sexy than in the weeks since January. Sometimes she wondered if she would ever be turned on again.

She watched Keir as he settled back into his seat, wrinkling his nose and looking across at Alan. 'We're getting outnumbered here, mate. Four of them . . . we never get a word in edgeways.'

'Say something worth listening to and we'll let you chip in,' Kath said. 'Not politics, though, I get enough of that at home.'

'You and me both,' Keir said. 'Which reminds me . . .' He was fishing in an inside pocket. 'The old man's got these tickets for a TV programme they're recording in Newcastle tonight. He got the tickets through the union, to take some of his mates. However . . .' He pulled a face. 'Since the steel strike folded he's been down on his knees facing east. So these are ours, if we want them.'

'How many?' Alan said.

'Six: one for everyone. We could catch the train through to Newcastle. It starts at seven.'

'What's it about . . . the programme, I mean?' Barbara was looking decidedly perky and Kath couldn't resist a dig.

'It's a talent contest, Ba-Ba. Now, all you have to do is brush up your yodelling and you'll be famous.'

But Barbara did not rise to the bait. 'Lovely, but I'd rather be notorious. Seriously, Keir, what programme is it?'

'It's some programme that discusses current affairs: they're debating the place of sport in the political arena. Should we ban sport with South Africa, should we boycott the Moscow Olympics . . . that sort of thing.'

'I think it's a disgrace that the Olympic team aren't obeying Mrs Thatcher,' Barbara said.

'Ah ha!' Keir was leaning forward. 'Now that's a Freudian slip, Barbara. "Aren't *obeying*" . . . so she's a dictator, is she?'

'You know what I mean,' Barbara said uneasily. 'She's our elected leader . . .'

'Not mine,' Keir said.

'Well, most of the people elected her . . .'

'Correction: she won in a first-past-the-post situation.

The *majority* voted against her or didn't vote at all. That's reality.'

'You're getting as bad as Jed,' Barbara said. 'He used to argue for argument's sake.'

'Apart from being unfair about Jed, that is the stock answer of the defeated opponent, Barbara. Come on, admit it, I've got you by the short and curlies.'

'Don't be disgusting,' Barbara said. 'Anyway, I only offered an opinion. I suppose that's a crime now, is it?'

'It will be if Thatcher has her way . . .'

'Shut up, Keir.' Alan had had enough. 'We see each other once in a blue moon now, so let's not waste time in wrangling. Who wants to go tonight?'

Only Kath and Elaine demurred.

'I can't come,' Elaine said. 'I'm . . . doing something else.'

That isn't true, Jenny thought. She doesn't want to come because she feels an outsider. 'Come, Elaine,' she urged. 'I'll call for you. It'll be a bit of fun. I've often wondered how they make a programme.'

'Honestly, Jen, I can't. I'm going home tonight . . . to see me mam.' Her eyes on Jenny were pleading to be let off the hook.

She really *doesn't* want to go, Jenny thought. Aloud she said, 'OK, I'll tell you all about it tomorrow.'

'What about you, Kath?' Keir was leaning forward to look up into Kath's face. 'Can't bear to leave the beloved for even one night?'

'There's a spare ticket if Elaine isn't coming,' Alan said. 'Bring Dave along.'

'Good idea,' Keir said. 'That'll make even numbers.' He turned to Jenny. 'That means I can be with you, Jen.'

Jenny tried not to glance at Elaine, fearful of what she might see, but the temptation was too great. She needn't have feared: Elaine was looking at Keir with an expression of complete indifference. 'She truly doesn't care,' Jenny thought.

'I wasn't joking in there,' Keir said when they were outside and making their goodbyes. He kept his voice low so as not to be overheard. 'I've missed you, Jen. And you haven't been too forthcoming on the phone.'

'I've been working,' she said, trying to edge back towards the others.

'So have I, but it hasn't stopped me thinking about you.'

Jenny couldn't think of anything to say – only the truth, that he had broken her heart. And that would mean betraying Elaine.

'I want to see you, Jen,' he went on. 'What about tomorrow? I'll ring you this afternoon.' His eyes were earnest under the thick lashes, the straight brows that were darker than the blond hair.

'I've got to work while I'm home, Keir. I'm awfully behind . . .'

'Join the club, Jen; do you think the rest of us aren't? But we've got to have some fun.'

Was it fun he wanted? Was that what she meant to him? He had put a restraining hand on her arm, forcing her to meet his eye. He looked so vulnerable, almost hurt by her refusal. Had Elaine got it wrong somehow? No, that was silly, you couldn't make a mistake like that. If it wasn't true, then it was a lie!

'Ring me tomorrow at tea-time,' she said, trying to smile. 'I'll see how much I get done through the day. Maybe we could get out for an hour . . .'

'Good. I'll ring at five, and please . . . please . . . be free.'

'Keep still, darling, I'm frightened I'll prick you.' Barbara stopped rotating on the stool and screwed her head round to look down at Jenny while her mother crawled round her, pinning up the hem of a new dress.

'I don't know why you worry about Elaine, Jenny. She may look fragile but she's as tough as old boots.'

Barbara's mother removed the last pin from her mouth and stuck it in and out of the fabric. 'That type of girl learns to cope early, they have to. Don't bother your head about her, Jenny. She'll survive.'

It was never easy to stand up to parents, especially someone else's, but sometimes it had to be done. 'I don't think Elaine is any different from Barbara and me, Mrs Finch. I don't think you would think so if you knew her.'

Mrs Finch's tailored eyebrows arched a little but her

mouth took on an even sweeter expression. 'I'm not meaning she's inferior to you or Barbara, Jenny, you mustn't misunderstand me. We can't help what we're born to, there's no disgrace in that.' She began to pick up more pins, with difficulty because of her long, pale-pink varnished nails. 'But she hasn't made the most of herself, as you and Barbara have done. I mean, what sort of future has she, with or without a baby? She'll never rise above herself . . . cafés and bars, domestic work. Whereas you and Barbara . . .' She cast a discreet glance upwards at her daughter.

'Well, you could do great things, if you stick in.'

She put in another pin and then returned to the attack. 'Look how hard you're working! Is Elaine capable of that? We have a very good education system in this country . . . or we did have till that awful Williams woman ruined it. If Elaine had had it in her she'd have been given her chance. Not that I'm criticizing her . . .' She gave a girlish laugh. 'We can't all be brain surgeons, can we? Some of us haven't got the brains. No, dear, you stop worrying – just turn a little, Barbara, darling . . . you stop worrying, Jenny, and concentrate on your studies. And when you've both passed with flying colours you'll have Jersey to look forward to.'

Jenny had been trying to formulate a counter-argument although with Mrs Finch in full flood it wasn't easy. But talk of Jersey was too much! 'About Jersey, I'm not sure . . .'

Mother and daughter both let out little howls of anguish. 'Don't say you're having second thoughts,' Mrs Finch said. 'Barbara's set her heart on it.'

'You promised, Jenny, ages ago. You promised! You were all for it.'

'I don't think I ever really said . . .' Courage failed her. 'I'm not saying I can't go, I'm just saying . . .' They were going to Jersey in July, and Elaine's baby was due in August. How could she leave her alone in the last few weeks?

'Well, don't worry about it now,' Mrs Finch said smoothly. 'You'll be keen enough when the time comes, when you've got all the exams behind you.' She leaned back and eyed her handiwork with satisfaction. 'You can get down now, darling. Mind the pins when you pull it over your head. I'll stitch it tonight, all ready to take back.'

She turned to Jenny. 'Has she shown you her dress for the Graduation Ball? Daddy nearly had a fit when he saw the bill but, as I told him, it's a once-in-a-lifetime affair.'

Jenny trailed dutifully upstairs to see the confection of silk-organza. 'It's hand-painted,' Barbara said, holding it against her. 'I'll need a terribly good strapless, won't I?' She chuckled. 'Or I could tape them up with Elastoplast like the models do.' She looked complacently down at her chest. 'Thank God I've got something to show. I used to think Elaine was repulsive, just skin and bone. The one good thing about the present situation is that she looks less like a candidate for Belsen.'

'I'm glad something pleases you,' Jenny said, feeling anger grow.

'She'll land herself a husband, you watch. She'll put on the little-girl-lost act – "I don't know how I got pregnant, your honour, it just grew" – and some poor sucker'll fall for it. As long as it's no one we know.'

Jenny thought of Euan's letter, in her handbag on the bed, and thanked several gods that she'd resisted the temptation to tell Barbara about it. *Tell me what she needs,* he'd written. *I'd help her anyway because I like her a lot, but you know how I felt about Jed. I'll do anything . . . just let me know.'*

She was interrupted by Barbara's arm around her. 'Wake up, Jenny Wren, you looked a mile away. I asked you what you were wearing tonight? We're sure to be seen on camera at some stage.'

Steven let the dog precede him through the yard gate and paused to latch it. 'In you go, bonny lad.' The dog turned, unwilling to go any further unless Steven was coming too. 'Go on, get in with you. It's like having a bloody shadow. What would you do if I went off the cliff?'

'Follow you, I expect,' Cissie said from the step. 'He's daft enough.' There was a smell of onion in the kitchen. Onion or leek, and a pan almost bouncing on the stove.

'Leek pudding?' Steven said hopefully.

107

'Maybe it is, maybe it isn't. Wash your hands and find out. And you sit down, Chips, before I stand on you.'

Steven stood at the sink to wash his hands and reached for the tea-towel.

'Not that, Steven, that's for pots. This is for hands.' The towel caught him round the neck and he wiped himself dry.

'The lad in?'

'He's upstairs. Working he says, but all I can hear is music. He says he's going through to Newcastle tonight . . . some tickets you gave him?'

'I got them off Stan Fairclough. It's a debate, Cissie, a TV debate. One or two members will be there and he needs to see how these things work. Like it or lump it, TV's the future. That's what dished Hume. It wasn't that he was a spineless bastard with a brain like a pea . . . he boasted he couldn't count without matches, boasted about it, but they didn't sack him for that. No, he got the chop because . . .' Here his voice became silky and his face took on a beatific smile. 'They sacked him because he didn't look pretty on the goggle-box. My God, what a country. Still, if that's the way the game's played I'll play it that way. I'm not proud.'

'Get that inside you then, and you'll be ready to tackle it.' She was scooping the wedge of suet pudding, heavy with leeks, on to a plate half-filled with crisp smoked bacon.

'By, this looks nice. Get the lad down. I bet he doesn't see the likes of this in Oxford.' Steven broke off a piece of bacon and passed it down to the dog, lying now between his feet.

'I should think not. It's my speciality this. And don't think I don't see what you're doing with that dog.' She went to the foot of the stairs that ran up the sitting-room wall and called, 'Grub up! Get yersel' down here.'

Keir came down with alacrity, sniffing the air like a bloodhound, planting a kiss on his mother's cheek in passing.

'Leeks and bacon, the food of the gods.' As he sat down opposite his father and took up knife and fork Steven's jaws ceased to masticate for a moment while he gloried in the fact that his son was not only clever and upright but handsome into the bargain.

Alan and Barbara and Jenny caught the train in Sunderland, joining the Belgate three who were already aboard. 'Do we get a chance to speak on the programme?' Barbara asked as they rattled through Seaburn and Bolton and on towards Newcastle.

'Don't ask me,' Keir said. 'It's supposed to be an open forum but you know TV – it'll all be stage-managed.'

They were met at the door of the theatre by a tall young man, disconcertingly like Antony Perkins, a badge on his chest naming him as 'Clive'. He took their tickets and waved them onwards. 'Grab a seat and a coffee. Somebody will be along to take care of you.'

The room was full of people, most of them looking bewildered. 'Do we get make-up?' a woman asked and looked downcast when Kath said, 'I shouldn't think so.'

From time to time girls in baggy jumpers and trousers, all with name-plates on their chests, flitted through the room calling out to unseen colleagues in what sounded like some strange and esoteric language.

'They're researchers,' Keir said, 'or production assistants.'

'There seem to be a lot of them.' Alan was looking disapproving. 'What will it cost to stage this for . . . how long? Half an hour?'

'They haven't spent much on the eats,' Kath's Dave said, discarding a cheese straw and looking mournfully at his plastic cup of coffee.

A girl in a turquoise jump-suit was bearing down on them. Her badge said 'Toni' and the outsize watch on her wrist was only slightly smaller than Big Ben.

'Right,' she said briskly, 'and you are . . .?'

'Keir Lockyer,' Keir said. Toni smiled briefly, letting her irritation show. 'I should've said "What are you?"'

'Male, single . . . ever-hopeful?' Keir suggested.

Her lips firmed but before she could speak Alan intervened. 'We're students, at least five of us are. Dave . . .' He gestured towards him, '. . . is a miner.'

The girl smiled brilliantly at Alan by way of thanks and turned her attention to Dave. 'Do you have strong views on the boycott?'

'Pardon?' Dave said, beginning to flush.

'The Olympic boycott,' Toni said patiently.

'Oh, that,' Dave said. 'It's up to them, isn't it?'

Toni's eyes glazed and she turned to Jenny. 'I think the athletes *should* stay away,' Jenny said, 'but we shouldn't make them. It's a free country.'

Suddenly Keir was speaking, his voice like a whip. 'What interests me is Mrs Thatcher's venom towards a country which is simply helping to maintain the legal government of a neighbour against massive outside interference in that country's affairs.'

Toni had brightened. 'But would you stop them going?'

'I'd appeal to them to uphold the Olympic tradition of brotherhood. That means they should go, whatever the politicians say . . . or do.'

The Antony Perkins lookalike was swooping down on them, running fingers through his hair like a man driven to frenzy.

'Oh God, Toni, the shit's hit the fan. The boom mike's down and Geoff is biting the carpet . . . literally biting.'

Toni appeared unmoved. 'Tell sound to get their act together and stop flapping, Clive. I'll be with you in a moment.'

'All right, all right.' He threw out his arms in one last despairing gesture. 'That's me out. I said it would be a shambles, I told Geoff, I positively grovelled to Nicola. Don't blame me when it all goes wrong.'

'Sorry about that,' Toni said smoothly, handing out tokens, white to Jenny and Keir, blue to the others. 'There's always a flap before we start recording. Don't let it throw you.' She waved Big Ben in the direction of a curtained door. 'Go through there and someone will show you your seats.'

A boy in denim was waiting beyond the curtains. 'Blue to the back, white to the front row,' he said, glancing at their tokens.

'Can't we sit together?' Barbara said.

'Sorry, blue to the back, white to the front,' he repeated, already looking beyond them to new arrivals.

'But we're together,' Barbara insisted.

'You can have my seat,' Jenny offered, terrified of the front row, unwilling to sit alone with Keir.

'No, she can't,' the boy said firmly. 'It's . . .'

'. . . blue to the back, white to the front,' Alan finished for him. 'Come on Barbara, you're holding things up.'

'They should've had Alan down here,' Keir said, when he and Jenny were settled on the wooden bench. 'He's got a tongue like a razor when he likes, but silky with it. That's his strong point. He'll cut a swathe when he gets to the Bar. I wouldn't like him prosecuting me.'

A man had suddenly appeared in front of them, standing beside a table bearing three microphones and three glasses of water.

'Ladies and gentlemen, thank you so much for coming tonight. We wouldn't have a programme without you. I'm Gwyllyn Govent, by the way. They tell me I'm the producer of this little event and I'm supposed to come out here and warm you all up so you'll be a wildly appreciative audience later on.'

He looked around with a self-deprecating smile on his face. 'He's done this a thousand times before,' Jenny thought. He was stroking his neck, feeling inside the rough cloth of his pink artisan's shirt.

'Now, I could tell you you've each been chosen because you're brilliant, articulate, witty, erudite – but you'd say, as the bishop says to the actress . . .' There was an appreciative titter. '. . . pull the other one. So I'm going to tell you the truth: that you've been chosen because we *hope* you're going to be brilliant, articulate, witty . . .' This time there was a roar of laughter and he looked suitably surprised.

He went on teasing them and they went on responding until a jolly good time was being had by all. 'I hope he's going to get on with it,' Jenny whispered but there was no response from Keir. She turned and saw that he was rapt, eyes fixed on Gwyllyn Govent, mouth curved in enjoyment.

'And now, ladies and gentlemen, I'm going to tell you a

111

little bit about tonight's recording. To be honest, we haven't a clue about what shape it's going to take. It's up to you to dictate the direction. You'll all get a chance to speak . . . if you hold up your hand someone will make sure there's a mike for you. Don't talk without a mike, you won't be heard. Don't talk over the top of someone else, neither of you will be heard. Don't say the F-word, you'll go to jail . . . and don't libel anyone or *I'll* go to jail. Otherwise you can say whatever you want. As long as it's relevant, that is.' He put up his hands and locked them behind his head, rocking backwards and forwards on the balls of his feet.

'I . . . think . . . that's . . . about . . . all. Jeremy, Nicola . . . anything to add?' No one spoke and then Toni suddenly materialized at Gwyllyn's side, whispering urgently.

'Shit,' he said. 'Shit, shit, shit.' He stormed out of the side door, leaping over the tangled mass of cables, and Toni turned to the audience.

'We have a slight problem with sound, I'm afraid. If you could all just relax, we'll come back to you as soon as we can.'

There was muttering and the sound of shuffling feet and then Barbara was there, in front of them, crouching down.

'I don't think much of the organization.'

'What organization?' Jenny said.

'Be fair.' Keir was leaning back, looking tolerant. 'It can't be easy, fixing up a complicated programme in a church hall. There are lots of technical reasons for a cock-up. Give them a chance.'

'Well, I hope there's something better than that awful coffee afterwards,' Barbara said, straightening up.

'Do you want to swop seats?' Jenny asked, but as she said it she felt Toni's eye on her. 'On second thoughts perhaps we'd better stay put.' Barbara looked disappointed but went back to her seat in the rows behind.

A few moments later Gwyllyn was back. 'Sorry about that . . . I think we're OK, now.' He rocked up and down again. 'Heads will roll if we're not. Now, where was I? Yes, don't speak without a mike, don't play to the camera, it's the presenter's job to address the viewers, and you speak to him. When we need you to applaud, that man there . . .' He gestured towards a denimed boy with a kamikaze

112

headband and a strained expression, '. . . will do this.' He held out his hands, palms up, and raised them to head height once or twice. 'Big applause please. We want it to sound emphatic. And keep on applauding at the end until we say stop. Right.' He rocked again. 'I think that's everything, so I'll now introduce you to our presenter for this evening. A big hand please for Mike Cattermole.'

'He looks much smaller in the flesh,' Keir said as they applauded the former newscaster. Mike Cattermole wore a navy shirt and tie, navy slacks and a cream jacket. His blond hair shone under the lights and his face had an unnatural orange hue.

He told a mildly obscene joke, chatted up a pretty girl in the front row, and then thrust his hand-mike towards a dungareed woman on the end seat.

'Have you got strong views on the boycott?' She launched into an angry diatribe but he was already withdrawing.

'We've got a lady of spirit here, ladies and gentlemen. If everyone feels as strongly as her we're in for an interesting evening. Now, in a moment we'll start rolling. I'll say a few words of greeting then I'll introduce our two "experts". Lots of clapping, please. They'll each have four minutes to put their case, and then the floor is yours.'

Beside her, Keir was almost trembling. 'He's enjoying it,' Jenny thought. 'He's excited by it.' The next moment a voice was counting down and the floor manager was leaping about, making gestures. The experts entered, bowed to the applause and sat at the table. One sipped nervously at his glass of water, the other beamed left and right, obviously hoping to curry favour.

'Good evening, ladies and gentlemen,' Mike Cattermole said . . . and then the experts were putting their opposing views.

'Thank you, gentlemen,' Mike said finally. 'And now we'll hear from the floor.'

Keir's hand was in the air and a mike was swinging down from somewhere above and behind him.

'Mr Chairman, I'm grateful to our learned friends here for giving us the pros and cons, so I feel somewhat guilty about saying I think they've both completely missed the point.'

All eyes were on him now, a camera was rolling noiselessly towards them and Jenny tried to shrink into the background. Keir finished speaking, and there was a round of applause. Someone else made a point and then Jenny recognized Alan's voice, reasoned and firm, weighing the advantages and disadvantages of sanctions of any kind.

Beside her Keir moved impatiently, his hand above his head once more. 'President Carter may have the American Olympics Association in the palm of his hand . . . I'm glad to hear he's succeeding at something . . . but this is Britain. We may not all be world-class athletes but we value our right to choose . . .'

Out of the corner of her eye Jenny saw Toni clutch her clipboard to her chest in a gesture of approval as Keir continued.

'We criticize Russia – with good reason, sometimes. But I don't think that a world power which meddles in every sphere, particularly in its own backyard, the Latin-American states, has any right to criticize another world power which simply means to intercede in a neighbour state, with entirely peaceful intention.'

'I don't agree with what he says but he's saying it brilliantly,' Jenny thought. And then, after what seemed like no more than ten minutes, the presenter was thanking everyone and his grandmother, the floor manager was wafting his hands to stimulate applause, and someone was shouting, 'Thank you. Wrap it up. Thank you, everybody. Lovely programme.'

They rejoined the others in the outer room and accepted a glass of pinkish wine.

'Jed would've *loved* that,' Kath said.

'It would have been a lot more anarchic if Jed had been there,' Alan replied. He turned to Keir. 'True?'

There was no answer. Keir's eyes were fixed on Toni, as she moved from group to group, thanking everyone profusely and asking some of them for details of their travelling expenses. At last she reached them.

'You were all wonderful,' she said, but she was looking at Keir.

'Thanks very much,' Kath said. 'Seeing that I never opened my mouth, I'm grateful for those kind words.'

'Well, you know what I mean,' Toni said. 'Not everyone can speak but you were a *lovely* audience.' She moved away, and Keir followed her.

'I wondered if I could ask you . . .' Jenny heard him say, and then he and the researcher were out of earshot.

'He was good,' Barbara murmured.

'He was *very* good,' Kath countered, 'but I wonder how much of what he said was meant.'

'What are you getting at?' Barbara asked.

'He sounded just like his father,' Kath said. 'That's what I mean. A bit posher, mebbe; more long words; but that was Steven in there tonight and Keir was only his dummy.'

Jenny sipped her wine. She thought Kath was being unfair, but she wasn't really concerned. It was far more important to her to know whether or not Elaine had lied about Keir fathering her baby. Perhaps tomorrow when he rang her, she would be able to find out.

11

16 July 1980

Jenny looked around her room for the last time, checking to make sure she'd left nothing except clothes that simply would not fit into her suitcase or holdall and must therefore be left behind.

In two hours' time she would be in a plane bound for St Helier, in Jersey.

'And I don't want to go,' she thought for the hundredth time.

The last few months had passed in a blur of work and waking in the night in a sweat of fear about her finals. In the end she had been too tired to resist Barbara's constant phone demands to plan the Jersey holiday; had even come to welcome it as she would have welcomed Hell if it had got her away from Liverpool. She had got a First, but was still not sure it had been worth it.

She picked up her bags and carried them on to the landing. 'I'm almost ready,' she called, knowing her father was prowling below fretting about the time.

Back in the bedroom she walked to the window. In a few weeks' time she was going to Coatham and life would begin in earnest, so the theory went. If only she could have had the next two weeks at home, to think, to plan . . . to look after Elaine in the final weeks of her pregnancy.

She leaned her forehead against the glass and acknowledged her self-deceit. She wanted to stay here in case she met up with Keir, that was the truth of it. In spite of his breaking his promise to ring her at Easter, in spite of all she had heard from Kath and Alan about his affair with the TV researcher they had met that night in Newcastle, in spite of the fact that he had fathered Elaine's baby as casually as though he had been pollinating a flower. None of it would matter two pins if only she could see him again.

'Jenny . . . you'll miss the plane.' She stayed still for a moment, trying to pretend she hadn't heard, then she bent to pick up her bags.

Her father was loading them into the car when Dave's car drew up beside them.

'I'm glad I caught you, Jen.' Kath looked happy and healthy, her skin caught by the sun, her hair streaked and falling untidily around her face. Jenny waved to Dave, standing by the car smiling, then turned to her friend.

'Oh Kath, I wish you were coming too.'

'Haven't I suffered enough? I've seen Barbara every day, almost, for three years and I've only just got rid of her. Have mercy! With a bit of luck she'll find herself a million-aire and leave you in peace.' Kath moved closer and clutched Jenny's arm, lowering her voice to a whisper. 'I've got something to tell you.'

Jenny put out an urgent hand. 'Before you tell me, promise me you'll take care of Elaine. I saw her last night – she's OK, but she's lonely since she stopped working. She'd like to have stayed on at the café but they didn't want her once she was really looking pregnant. Tell her I'll come as soon as I'm back. It's only two weeks.'

'OK, OK, you know I'll look after her. Now listen to me.' Kath's eyes were shining as she looked swiftly back to Dave, and then murmured, 'I'm pregnant too, Jenny. I only found out today, and I'm not sorry.'

'Kath . . . are you sure?'

'Yes, I'm positive. Now that I know for sure, I'm pleased . . . and Dave is over the moon.'

'Oh Kath . . . as long as it's what you want. What will your mother say? But if you're getting married . . .'

Kath's face dropped a little. 'That's a bit tricky. Still, I know me mam, she'll come round in the end, when she sees it's going to be all right. And *of course* I'm getting married, and you're my bridesmaid!'

Keir waited until the morning paper had been read, partly digested and then thrown down by the fender. 'One and a

117

half million out of work – one and a half million! She'll have this country bankrupt before she's finished.' His father's face was gargoyled in fury and Keir was careful not to offer any counter-arguments.

After a while there was a noisy searching for boots and parka, the dog bright-eyed and quivering by the back door. 'Come on, then,' Steven said at last and Keir made his move.

'Going down the sea-front? I'll come with you.'

They walked down the street and along the side of the allotments, the dog gambolling ahead, looking every now and then to make sure they were there.

'He's come on, hasn't he?' Keir said and saw his father's cheeks suck momentarily inward in satisfaction.

'Aye, he's got a year or two left yet.'

'Did you ever find out who he belonged to?'

'No. Your mother told the pollis we had him. "If anyone turns up, send them round," she said but we wouldn't't've given him back. They didn't deserve him.'

'He might've wandered off. You couldn't blame them for that.'

'No,' his father said equably, 'but I can blame them for not walking their shitting legs off till they found him. I can blame them for that, all right.'

They had almost reached the sea now. It lay ahead, blue and sparkling, red lobster pots bouncing here and there far out from the beach.

'I've been thinking, dad – I can't just sit around here waiting for a job. I know you've got feelers out, but what if nothing comes of it?'

'Something'll turn up.'

Keir sought desperately for a clinching argument. 'You think you have friends, dad, but you've always been outspoken, and you'll have enemies too. They could put a block on.'

'I've no doubt I've got enemies – and in high places. It's always the scum that rises to the top of broth. But you should have a bit more faith in yourself, lad, and in what you've got to offer. They'll take you. They'll jump at the chance, but you'll have to be patient. Jobs don't fall vacant just to order.'

Keir wanted to turn to his father and state the bald truth, that with every day that passed he grew less and less enthusiastic about politics. It was too slow, even for those who succeeded. Five years of lip-service – at least five years – before you stood a chance of nomination. He had looked around at ward meetings and seen other contenders. If he got a job with the party or the unions it would shorten the process, but he would still be thirty before he even stood for election. Given a safe seat, and the way things were going that looked more and more likely, he might be in Parliament by 1988. But for what? For five years' further lip-service before he stood the least chance of office. He had seen the northern MPs catching their early Monday morning trains, grey men mostly, faceless and indistinguishable from the Tories, gazing out of first-class carriages as the train carried them on towards another boring week. Keir didn't know what he wanted out of life but he was becoming more and more certain that it wasn't that.

He thought of Toni, already a production assistant, making decisions, seeing her ideas translated into programmes. He was meeting her tonight in Darlington. Until then he might as well forget it. He had never been able to talk to his father.

May Denton watched the house opposite through the mesh of the lace curtain. There was a place in the pattern where she could see quite clearly and it was at eye-height which made it even easier.

She had taken to watching the house across the way since the new tenants had moved in, fascinated by their loud voices, their bright informal clothes and the swarm of cats and dogs that followed them everywhere, even into the car.

The woman had come across and introduced herself, right at the beginning. 'Do come over when we've settled,' she'd said. 'We're quite harmless.'

May had been polite and wished them well, but the next time she saw the woman coming up the path she had cringed out of sight, her heart thumping uncomfortably as the woman knocked and banged the letterbox.

119

Watching now as the woman came out with a jug to water her garden, May tried to remember if she had ever been sociable. Perhaps once, when she had been in service. She and her friends had laughed a lot then, just over little things – laughed and cried and gone to the pictures together. In the war she'd had friends, close friends even.

'Yes,' she decided, moving away from the window as the woman's front door closed behind her, yes, once she had welcomed people, but that was *before*. She felt the old fear start up inside her and rushed to quell it in the only way she knew how: tidying her kitchen, taking each item from the workbench to wipe it with a dishcloth and dry it with a tea-towel until it shone.

If she had stood up to them more, shut her ears to the things they said, would it all have been different? And Edward had died of a pulmonary embolism: she must remember that. It had been written on the certificate, right there in black and white, so how could anybody say he had died of shame?

To comfort herself May went through to the dining-room and opened the sideboard drawer to lift out the various boxes and cases it contained. The cutlery gleamed when it was revealed. Each item was polished on the first day of the month, unless it was a Sunday in which case she did it on the second. Now she counted each knife, fork and spoon and noted them on her pad before replacing them all and closing the drawer.

She was making an inventory of her possessions, trying to assess their worth: so much had been parted with over the years, things that Edward had prized. There was a sudden trembling in her chest, an awkward flutter that made her reach for the back of a chair.

If the worst came to the worst there was the house: that was worth something, more every day if she believed the cards in the estate agent's window. When she felt better she opened another drawer, full to the top with table linen. Edward's mother had had beautiful hands, and there had been an aunt, too, who had stitched and embroidered, even down to towels and traycloths. May embroidered herself, but not to the same high standard. She had loved to launder the linen, see it come up white and sparkling, the colours bright enough to leap out at you. She had treasured it all before it belonged to her. Long before.

She selected a cloth heavy with crocheted lace, and a set of place mats, anemones on an ecru background. Tomorrow she would take them to the market. The woman there was a lady, turning things in reverent hands before she mentioned a price.

The drawer still seemed full; there was no sign that anything was gone. Not that Alan ever seemed to notice things in the house – or if he noticed he forbore to comment.

Remembering how handsome he had looked when he went out this morning, May smiled, putting the crocheted cloth to her lips to hide her satisfaction. She had done as Edward wished: she had raised his son to be a gentleman. And he had always had good limbs, no matter what they said. How proud Edward would have been to see Alan now.

She should not have allowed herself to think of Edward. The picture came, the terrible picture of his face when the letter arrived – his poor, torn, grief-stricken face. And then the turning away from her, that had really been turning his face to the wall.

Sunlight was flooding the cabin now and below the clouds were pink. 'It's rather nice,' Jenny thought as the shore of Jersey came into view far below, looking disappointingly like England. There was a bump and a lurch and they were down, and dozens of little grounded aircraft seemed to be whizzing past the windows.

'Everyone here has a private plane,' Barbara said, and then they were out on the shaky steps, walking into the airport building, and anxiously looking to reclaim their bags.

'I'd go straight back home if my luggage was lost,' Barbara said. 'No point in coming to Jersey if you can't dress up.'

'See,' she said later, as their bags were safe at their feet and they were looking around them. 'This is where the money is.' Then a tanned man in a knitted T-shirt and a

blonde woman with pearls the size of hen's eggs in her
ears were bearing down on them.

'Barbara! And this must be Jenny. Come on, the car's
outside.' Aunt Bea smelled of every perfume counter of
every store in all the world. Her fragrance wafted them out
and into the long red car.

'It isn't very French,' Jenny thought and then a street
sign on a wall caught her eye: 'rue de Beaumont'. Jersey
was strangely countrified, with stone walls, and tiny indi-
vidual houses, and flowers everywhere you looked.

'You're going to have the time of your lives,' Aunt Bea
said, from the front seat, her pearl ear-rings quivering in
emphasis. 'Uncle Eric and I are quite determined. They
deserve it, don't they, Eric?' The bracelets on her wrists
clattered agreement. In the mirror Uncle Eric's eyes were
on Jenny, smiling as if to say, 'Humour her, she means
well.'

'It's awfully kind of you to have us,' Jenny said.

'You've changed your tune,' Barbara chuckled. And then
to her aunt and uncle: 'She didn't really want to come.
Mum and I had to twist her arm.'

'Did you think we were going to eat you?' Uncle Eric
said and Jenny, mortified, shrank into her seat and concen-
trated on the view until they reached the white-painted
guest-house, set high above St Helier, with 'The Dunelm.
Exclusive Guest-House' painted at the gate.

In their immaculate third-storey bedroom, Jenny
watched Barbara unpack with round eyes. What Barbara
had brought with her was a trousseau, dresses of every
colour with matching sandals, wispy waist slips with
matching briefs, bikinis in sharp jewel colours with toning
jackets. When it was all disclosed she looked at her own
case.

'Come on, Jenny,' Barbara said. 'Get everything hung
up. I can't wait to go out.' From the window where she
stood Jenny could see a turquoise sea sprinkled with jagged
rocks, and tiny white boats weaving in and out. Below, the
garden was full of pink lily-like flowers.

'I expect they're Jersey lilies,' Barbara said.

'I thought that was Lily Langtry?'

'Oh well, I only suggested. That's buddleia there, I do

know that. And hydrangeas. Daddy came back green with envy last time he was here. You know how he loves his garden, but he can't compete with this. Come *on*, Jenny, I want to go sight-seeing.'

Barbara's enthusiasm evaporated a little after ten minutes in the street. 'It's a bit like Blackpool, isn't it?'

'What d'you mean?'

'Well, all the . . . well, ordinary people.'

'What did you expect, Ba-ba? We're still in the British Isles. Anyway, these *are* ordinary people, if I'm any judge. Holiday-makers, half of them from County Durham probably.'

'Oh God,' Barbara said, 'I hope not.'

She cheered up a little at the sight of a harbour choked with yachts. 'The money there must be here,' she said, lifting her gilt-trimmed sunglasses to gaze on the forest of masts.

Behind them a cliff rose up sheer, and when they explored the maze of shops beneath they found every second one, or so it seemed, sold perfume.

'I'm going to take back stacks,' Barbara said.

'You're only allowed so much.'

Barbara gave Jenny a knowing look. 'Be your age, Jenny. I didn't say I was going to declare it.'

They turned at last and began to walk back towards the guest-house, between buildings both quaint and ugly, crowded together. Past a statue of Queen Victoria, green as grass and plump as a pouter pigeon, along streets named 'Rue' and 'Le Petit' with English names superimposed. 'This island hasn't made up its mind,' Jenny thought and felt an odd sense of kinship.

She could never make up her mind about allegiance. Look how she felt about Barbara, sometimes loathing her, at other times feeling a surge of affection for all the good times they had shared as children. She remembered Barbara, pig-tailed and determined, pushing a boy who had bumped into Jenny and made her cry. 'I must try to make this a happy holiday,' she thought. After all, if Kath could endure three years of Barbara she could put up with two weeks.

Back at the guest-house they ate at a table in the dining-room, served by Irish colleens with creamy complexions and thick legs.

'Everything all right?' Aunt Bea called from time to time. Uncle Eric moved about with bottles of wine, opening them with a flourish and standing, head on one side, until the guest signified approval.

'We'll go out again as soon as we're finished,' Barbara said, spooning in crème caramel as though it were medicine that had to be got down quickly. 'I want to spot the best places to socialize at night.'

Jenny suddenly rebelled. 'I'm sorry, Barbara, but I want a rest. I will come out but I want half-an-hour's peace . . . with my feet up, if possible.'

They lay on their respective beds while the light faded and lights sprang up around the bay. The noise from the streets drifted up, more loudly as the evening wore on. Jenny tried to do a crossword but Barbara's sighs grew louder and more aggrieved until at last she put aside the paper and swung her feet to the floor.

'OK, I give in. Where are we going?'

'I've got a list,' Barbara said, suddenly animated.

'I thought you might have,' Jenny said and made for the bathroom, trying not to wish that it was Kath, so recently arrived in her life, who was with her now. And then she remembered Kath's news and felt a flutter of apprehension: in spite of Kath's brave face there were bound to be problems. It was one more thing they hadn't planned for; another reason to suspect that that night on the hill had bestowed not a blessing but a jinx.

'Mam.' Kath stood in the doorway, trying to think of what to do with her hands.

'What is it, pet?' Her mother was in her chair by the fireplace, her bare feet resting on top of her fluffy slippers, discarded because of the summer heat.

'You should have some sandals for this weather,' Kath said, moving further into the room.

'Is that all?' Her mother smiled. 'The way you said "mam" sounded like the crack of doom.'

'It is,' Kath said.

There was a moment's silence while her mother fumbled her feet back into her slippers as though to gain strength from them.

'What do you mean?' she said at last and then, 'as if I didn't know.'

'I'm sorry, mam, I'm really sorry. We didn't plan it this way.'

Her mother was leaning forward to lift the kettle on to the glowing coals.

'Well, never mind. Let's sort out what's for the best. Don't tell your father.'

'He'll have to know some time.'

Her mother looked up, her face suddenly contorted with fear. 'You're not going to keep it?'

'Of course I am, mam. I want it . . . now that I've got used to the idea. And Dave's like a dog with two tails.'

'The sod will be, now he's had his way.'

Kath had never heard her mother speak with such venom before and it shocked her. 'Don't be like that, mam. I know it's too quick, but it would've happened sooner or later anyway. We're getting married – we love each other. I thought you'd be pleased.'

'Pleased?' Her mother was standing now, seeming suddenly taller and more menacing than Kath would have believed possible. 'Pleased to see you toss your life away on a lad who has no more sense than to drag you down to his level? Pleased to see you marry a pitman and drudge your life away washing pit clays? Pleased to see every hope, every dream I had for you, gone in one bloody minute of cock?'

'You're frightening me,' Kath said, moving backwards. She had never heard her mother use obscenities before. Swear at a leaky washer or a cake that failed to rise – yes. But this was different.

'Frighten you . . . I should do more than frighten you, but I'm scared of what I might do if I let go. Get out of my sight! Go on, get out, get to hell out of it before I do something I'll . . .'

Kath had expected a brief outburst of anger, fury even. But it should have been over in seconds. By now her

125

mother's arms should have been around her, comforting, pledging solidarity. *'We'll work something out, Kath. Leave it to me,'* – that was what she had expected her mother to say. Not, 'Bitch, bitch, silly little bitch. You've ruined everything now.'

For the first time Kath knew doubt. She had wanted the baby almost from the moment she had realized its exist- ence, but not at this price.

'Please, mam . . . please, just listen to me. I've got it all planned. You'll love Dave when you really get to know . . .'

She never finished her plea. Her mother's hand was suddenly upraised as though to strike her.

'Don't mention his name, don't say it in this house or I'll kill you . . . I'll kill you.'

Her voice broke then and she subsided, weeping, into her chair. But when Kath made to touch her she fended her off. 'Don't touch me, I can't bear you to touch me. Do what I told you . . . get out of my sight.'

'I love him, mam.'

'You love him now, I dare say you do.' She got up, went to the draining-board and began drying the stacked dishes. 'And you'll go on loving him for a bit, while he's still taken up with you, still dancing attendance. And then you'll have another kid, and another, and maybe a fourth and fifth. He'll be off down the pub every night and when he isn't he'll be in the chair asleep, knackered, stinking of sweat because he's hewed coal all day.' Her voice was rising and her hand inside the brightly coloured tea-towel went round and round a single plate until cloth squeaked against glaze. 'And gradually . . . just gradually . . . you'll feel yourself going to pot. You don't dress up any more. Why should you? The only time he ponces himself up is for a night out with his mates. After a while you can hardly be bothered to get washed. And then . . . and this is the worst part . . . you feel your brain going. You can actually hear it setting in your head, going solid like a jelly, till all you can think of is buying sugar and tea and marge, and putting your club money by. That's what I wanted to save you from. If she can have her chance, I

126

thought, if she breaks free I won't mind. And you've thrown it all away.'

'Jenny will be in Jersey now,' Elaine said. Across the table Alan nodded. 'I don't think she wanted to go in the end,' Elaine went on, grinning over the rim of her lager and lime.

'No,' Alan said, thinking of Jenny's earnest strictures about looking after Elaine while she was away, 'no, I think she'd rather have stayed at home.'

The pub was quiet this early in the evening, only the landlord behind the bar and a solitary customer on a stool.

'Have you heard from Euan?' Alan asked.

Elaine nodded. 'He writes about every two weeks.' She flushed a little and bit her lip. 'He sent me some money, as a matter of fact. A cheque.' It was there in her bag, folded and a bit grubby. 'I haven't cashed it, I want to send it back.'

'He'll be hurt if you do. He was very fond of Jed . . . and you. You know that. Now that things are as they are, he wants to help. We all do. And Euan can afford it. I think you should cash it and tell him what you spent it on. He's a funny guy. He loves details. "Tell me again," he'll say, about crazy little things.'

'I know what you mean.' Elaine was smiling but Alan could see she was still troubled. Why didn't she want to use Euan's cash? God knows she must need it. All the same, she was looking well, her cheeks rounded, her mouth somehow softer, the huge dark eyes less shadowed than they had been.

'Have you made any plans, Elaine? For afterwards, I mean? I know Jenny's coming back to be with you when you have the baby . . . but after that we'll all be going our separate ways and I don't like to think of you here on your own.'

She was looking astonished and grateful that he cared about her future, and it made Alan angry. He had seen that look before, on his mother's face: surprise that anyone should take account of her.

'We really do care about you, Elaine, all of us. Partly for

127

Jed's sake, I'll admit, but also because we like you. You were one of the gang. Not for long, but neither was Euan, and you know how we care for that old sod. This baby is all we'll have left of Jed . . .'

Elaine was crying now, quietly and without movement, tears welling up and running down thin cheeks, long eyelashes sticking together to make her eyes look wider and darker than ever.

'I'm sorry, Elaine. I was trying to be tactful but I messed it up.' She was shaking her head and scrabbling in her bag at the same time. He reached for his own folded handkerchief, thinking that was what she was searching for, but when her hand emerged he saw she was holding two pound coins.

'Will you get the drinks in? Same again for me, please?' He was going to refuse her money, knowing he had just enough for another round in his pocket, but then he realized that would be inept.

'Of course,' he said, taking the coins from her bitten fingers. 'And when I get back remind me to tell you a funny story I heard today.'

12

22 July 1980

It was half-past eight but Kath lay on in bed. She disliked getting up now, and going downstairs into the icy atmosphere of her mother's anger. The old comradeship between them had vanished as if it had never existed. Now her mother spoke only if it was essential, and all little acts of affection had ceased.

It was awful to live with wrath, but worst of all was the realization of how much her mother had hated her life in a pit village. 'All these years,' Kath thought, 'all these years, the only thing that's kept her going is what she hoped for for me. And now she hasn't even got that.'

In the beginning, when the hoped-for period had failed to arrive, Kath had wanted to sit down and cry. And then she had seen the look of satisfaction, almost of joy, on Dave's face when she told him, and suddenly it had seemed all right, sensible even, the logical thing to happen between two people who loved one another. Her mother's reaction had brought her sharply back to earth and for a moment she had even wondered whether or not to go ahead with the pregnancy. And then she had talked again to Dave.

'She can't live our lives for us, Kath. She's your mam and I respect her for that, but we have to decide for ourselves, and if it's what we want that's it.'

'If only it was that easy,' Kath thought now, reluctantly pulling herself up in the bed. There was no sign of the cold war ending, and although she sensed that both father and brothers were sympathetic to her, they were too afraid of her mother to show it.

'You're on your own, kid,' she told her reflection in the bathroom mirror – and then she thought of Dave, patient, solid, there when she needed him. It was enough to bring a smile to her face, a smile that lasted until she was downstairs and moving towards the kettle on the hob.

'Any tea?'

'If you make it.' Her mother's voice was indifferent.

'Do you want a cup?'

'No.'

'You could say "no, thank you".' There was no reply and Kath felt her eyes prick. She moved the half-full kettle on to the fire and turned back to her mother.

'Come on, mam, we can't go on like this. It's childish.' Still there was no answer and she moved forward, putting a hand on her mother's arm. 'Come on, I've said I'm sorry. I've said we're getting married but I'm still going to teach. Dave's a nice lad . . .'

Her mother withdrew her arm, flinching as she did so, as though Kath's very touch was offensive to her.

'Oh, mam,' Kath said but when she made to take the weeping woman in her arms she met with resistance.

'Do you think you can make it right with a cup of tea and a cuddle?' Her mother's voice was icy. 'Do you think you can give me back twenty years of effort in one minute's consolation?' She moved towards the door, like a very old woman. 'I'm going up to lie down. I don't want anything . . . except to be left alone.'

They had been in Jersey for a week now, sitting in the sun on almost deserted beaches, driving to tiny coves where trees ran right down to the water, eating lobster as though it were cod, and drinking more than was good for them.

'I'd love to live here,' Barbara said, as she drove towards St Brelade's Bay, the back seat piled with swimming gear and packed lunches and all the paraphernalia of a day at the seaside. 'Look at that house – isn't it wonderful? And *detached*.'

'It looks pretty well anchored to me,' Jenny said.

Barbara looked vacant for a moment and then caught Jenny's meaning. 'Oh yes, very clever. You know what I mean. I'm never living in a semi again once I leave home.

130

And I want stained-glass windows like that one, and a circular drive.'

There was a tiny red-sailed yacht out on the turquoise sea, and geraniums and blue hydrangeas were spilling from the gardens as they drove down La Maquandarie towards the bay. Flamboyant pink lilies showed in every garden, their trumpets as big as tea-cups.

'Do you know what puzzles me?' Barbara asked as they parked by the almost deserted beach.

'What?'

'Well, we never, or almost never, see anyone sunbathing, but they're all brown.'

'Perhaps they do it at night?' Jenny was smiling as she said this, acknowledging that seven days of living with Barbara had reduced her conversation to a very basic level. But Barbara, already scanning the horizon for talent, made no reply.

They lay on the warm sand, turning occasionally like joints on a spit to get an even roasting.

'I could stay here forever,' Barbara said. 'In fact . . .' She sat up suddenly, seeing a red sports car pull up above the beach. Jenny had seen the look on her face often in the last few days. Barbara eyed Jersey rather as a hungry man might eye a side of beef.

The man coming down the beach was tall and fair-haired. 'I bet he's a pilot,' Barbara said. 'Jersey's always full of air crew.'

'More likely he's an articled clerk from Luton over here for his hols,' Jenny said and rolled over on to her face.

After a while she sensed that Barbara had lain down again. 'Has he gone?' she murmured against her sandy arms.

'He's in the sea. He's left his clothes quite near us.'

'Is he doing a Reginald Perrin?' Jenny asked.

'A what?'

'You know, walking into the sea never to be heard of again.'

'Don't be silly. Actually, he swims very well. I think I'll go in.'

'Want company?'

'No. Unless you want to . . .'

For a moment Jenny was tempted to get up and accompany Barbara into the sea just to queer her pitch with the sports-car owner, but the warmth of the sand was too seductive. 'No, I'll stay here. Don't get wet.'

'No, I won't . . . what do you mean? Honestly, Jenny, sometimes you're completely puerile.'

When she had gone Jenny rolled on to her back, the sun making purple patterns inside her closed eyelids. Perhaps Barbara would feign drowning and be carried lifeless from the sea by a bronzed lifeguard? Perhaps the sports-car man was really Lloyd Bridges with a snorkel concealed in his trunks? She sat up but could only see two heads bobbing in the waves, a few yards between them.

When she looked again Barbara was striding from the surf, trying to look like Ursula Andress, the sports-car man in close attendance.

'Oh no,' Jenny said under her breath, and then brightened suddenly at the thought of having her remaining time in Jersey to herself.

In fact the man, who predictably was called Guy, was not a pilot but a solicitor from England working for a firm in St Helier. 'Do you come from Luton?' Jenny enquired sweetly and received a warning glare from Barbara.

They walked up to the cars and changed out of their swim-wear in their respective back seats before driving to an inn high above the bay for drinks. The girls sat in the garden while Guy went inside for three glasses of lemonade shandy.

'Isn't he gorgeous?' Barbara said, repairing her make-up with the aid of a small handbag mirror. 'Do I look all right? His mother's a magistrate.'

'How, in God's name, did you manage to find that out? You've only known him five minutes.'

'We were talking about law and order . . .'

'In the sea?' Jenny said weakly.

'Well, as we were coming out. I said Jersey was so peaceful . . . he just said it casually. He wasn't boasting.'

When Guy came back he seemed pleasant enough, dividing his attentions equally between the two girls, getting up to replenish the drinks and waving aside offers to pay.

Jenny was becoming fascinated by the couple at the next table. The man had his back to her but the woman facing her was beautiful in a sultry way, with black hair falling on to her shoulders. Wasps were flying around the tables, settling on empty glasses or scraps of food, and one, reconnoitring the woman's glass, tumbled in. It struggled fiercely to free itself from the surface of the water. The man reached to help it but the woman stopped him with a beringed hand.

The wasp's struggle seemed to go on for hours. It threshed about in the lager while Jenny sat, craven, wanting to get up and bring its agony to an end but afraid to interfere. Watching, the woman's face grew catlike. 'She's enjoying it,' Jenny thought. The surface of the liquid suddenly began to vibrate fiercely and then, just as suddenly, was still. The man had shifted uneasily in his seat once or twice. Now he signalled to a waitress to take the glass. She picked it up and slung the lager on to the nearby garden.

'Please let it be all right,' Jenny thought but she knew the wasp was dead. The man and woman stood up and moved away and something in the set of the man's shoulders showed he had not enjoyed the spectacle. 'I hope he gives her up,' Jenny said and realized that she had spoken aloud.

'I beg your pardon?' Guy's eyes on her were polite and enquiring.

'Oh, nothing' she said. 'Is it time to go?'

'Do you think there's enough?' Cissie asked. Steven eyed the laden table and sniffed the aroma of sausage rolls warming in the oven. 'Enough? Cissie, pet, he's bringing one girl home. One girl! And if she's a London girl she's likely afraid to eat anyway.'

'I want everything nice.' She was looking around the living-room, dissatisfaction on her face, and Steven's eyes narrowed.

'What's the matter, Cissie? Place not good enough for company? It's our lad she fancies, mind on. If it was good

133

enough for him to grow up in, it ought to be good enough for her.'

'Shut up, Steven, we don't want your soap-box this afternoon. And fasten your collar and . . . my God . . . get something on your feet.'

'I am, I am.' He sat down, grumbling, the dog putting a sympathetic head on his knee and complicating the business of the boots.

'They're here, Steven, I can hear a car. You know she's got a car.' But the car did not stop at the back door, and another ten minutes passed before Keir was shepherding a small, made-up girl through the front door. Steven was just about to ask why he hadn't come in the usual way when Cissie's elbow caught him painfully in the ribs.

The girl was affable and friendly – 'far ower friendly,' Steven told himself as he sat with an unfamiliar cup and saucer, listening to conversations about 'shoots' and 'recces' which, when he demanded an explanation, turned out to be filming and nosing around for something to film.

'Do you like television?', the girl asked him, obviously trying to draw him in to the conversation.

'Not at the moment,' he said. 'Too much Olympics. Ovett and Coe, Coe and Ovett. What a pair of prima donnas.'

'And Daley Thompson,' Keir said. 'He's a cinch for a medal.' Steven could see the young pair warming to a subject that didn't suit him, and decided to create a diversion.

'Did you hear the news at dinner-time? Nearly two million unemployed. Worst figures since '36.' He stuck out his chin at Toni. 'And all your lot can talk about's some poncy athletes. There's lads down the pit could run rings round them, *and* after an eight-hour shift.'

'Dad thinks miners are the salt of the earth,' Keir said. There was a strange, almost rueful note in his voice that Steven couldn't fathom.

'It's a dying industry, though, isn't it?' Toni said. She turned to Steven. 'I know it's a traditional way of life in the north-east, but you can't expect future generations to accept it.' She was looking hard at him and again he got the feeling of an undercurrent.

'Are you working up here?' he said, ignoring what she had said about the pit.

'Don't be nosy, Steven,' Cissie said, wafting the teapot about like a flag. He held out his cup for a refill.

'I'm only asking a simple question, Cissie.' He turned back to Toni, willing an answer, but she had looked to Keir, obviously waiting for him to speak. He put down his cup and braced his shoulders.

'Toni's come for me, dad.' He turned to his mother. 'I'm going to London, mam. I've got a chance of a job in a film-production company. Just a small concern but I've got to start somewhere if I want to get into television.'

Steven couldn't speak but his mind was whirling back and forth. *London? TV?*

'What will you be doing?' Cissie was saying, round-eyed.

'I'll be a runner, at first. A dogsbody. But I'll learn the trade, and then I can apply for researcher's jobs at Eastern. Toni's with Eastern Television. We want to work together eventually.'

Inside Steven seething emotions were beginning to settle. It might not be that bad: all anybody ever thought about nowadays was television. It could be a bonus – a degree and a job in television. If he married the right girl . . . Steven looked at Toni, at the patched jeans, the checked shirt more suited to a navvy, with tits showing through. That would go down well with a selection committee! Like a load of lead shot!

'When d'you plan all this then?' She was putting a hand on Keir's thigh, as though she owned it, cheeky bitch.

'We're going today, this afternoon, dad. Toni has three days off to help me settle.'

'Where will you stay?' He knew the answer before he asked, and the quick glance between them only confirmed it. He let it go for fear of upsetting Cissie who was leaping around bleating about clean shirts and underpants. 'What happens if the job doesn't work out?'

'It'll work out all right, dad. It's what I want. I knew it the night I met Toni . . . that night you got me the tickets, remember?'

135

When they had driven off in the posh car, the back seat piled with all the stuff that had just come back from the college, or so it seemed, Steven went out to his pigeons.

He had meant to cull the old blue cockbird today but he hadn't the heart. It cocked its head and gave him a beady red eye. Behind it, its offspring cooed and hustled for the corn he threw.

'It's a bugger, isn't it?' he said to the bird. 'You give them the best, you make sacrifices, you play it by the book. And then they shite on you.' He threw the last of the corn and turned back to the step, feeling the dog warm against his leg. The worst of it was that he had given the lad the tickets himself and started the whole ruddy ball rolling. That's what you got for being clever. He didn't know much about television but he had an uncomfortable feeling that it was no job for a Durham man.

If the lad kept his nose clean in London it might work out – but London was a bloody sink of iniquity. And the girl had looked a right little madam.

Cissie came into the yard and looked at him. 'What's the matter with you then? You look as though you could walk under a duck.'

He was about to give her a mouthful when he saw her eyes were wet. He went through the door, shutting it with one hand, scooping her into the other arm. The dog got caught in the door and yelped itself free.

'Come here, you silly faggot. He's only gone to London, you know. He'll be back before you can say Jack Robinson.'

'No, he won't,' Cissie said, fishing in her cardigan pocket for a hankie. She blew her nose and then looked up at him. 'He won't be back because she won't let him. She's got her hooks into him. I could tell.'

Steven wondered if he should tell her that it was a long time since Keir had been wet behind the ears, in that direction anyway, but decided against it.

'She won't last a twelve-month, Cissie. I've seen it all before. Let him have his head with this daft job. What was it – a "runner"? By Christ, I didn't think you needed to go to Oxford for that . . . Still, let him try it and then he'll come scuttling home. Now, get the kettle on and make a proper cup of tea – in a mug, mind. I've had enough of the lah-di-dahs for one day.'

Jenny had left Barbara behind in a welter of rejected clothes, tonging her still damp hair, plucking anxiously at an imaginary spot. 'I hope you don't think I'm letting you down, Jenny. I wouldn't ordinarily, but this is it! This is *the* one. We have so much in common, I can't believe it.'

So Jenny had caught a bus to the zoo alone and now she wandered from enclosure to enclosure, seeing the cheetahs lying in the shade, a beautiful, spotted heap; seeing a snake's cast-off skin dangle from a tree like a bag of doubloons, and little orange monkeys swing fireball-like through the branches.

A man beside her said, 'This is the best zoo in the world – seven hundred different creatures,' but to Jenny the eyes were all the same, age-old and sad.

She watched two female monkeys groom a complacent male, their tiny fingers picking and plucking. Somewhere was Barbara doing that for Guy? She laughed out loud and had to scurry away when a woman beside her looked askance.

She found herself back with the tiny monkeys. Close up they were like agile Pekinese, two of them playing with a feather, chasing and snatching like children. She stood for a long time, remembering childhood freedom, the days when finding a feather was all that was needed for bliss. She had waited such a long time to grow up – now she sometimes wondered if it was worth it.

She walked down to the lake, seeing flamingoes pink and naked-looking until their wings flapped to reveal scarlet and black, then back towards the monkey-house again, past a sign asking politely for quietness. Behind glass the monkeys sat, disdainful, while two human children dashed up and down the aisle making as much noise as they could in spite of the notice. Jenny waited for their parents to remonstrate with them, but they simply smiled indulgently until the monkeys were driven to pick up their offspring and lope out of sight.

137

'I shouldn't feel this, but I do,' Jenny thought. 'Animals are superior to humans.'

The joy of the zoo lasted all the way home, and until she had bathed and changed and gone out on the town with Barbara.

'To Jenny and Barbara,' Guy said. They clinked their glasses in return and sipped their snowballs.

'Umm, lovely,' Barbara said. They were sitting on stools at the padded bar. Behind them the floor area was dark except for a circling light on the ceiling that flashed eerily from face to face as the beam passed. On the stage a small brunette was singing 'La vie en rose' after the manner of Piaf, a three-piece combo behind her providing the musical accompaniment.

'It's not like France, is it?' Barbara said.

'You know France well?' Guy looked suitably impressed.

'We go almost every year,' Barbara said, looking Guy squarely in the eye. To Jenny's certain knowledge Barbara had been to France three times, once with the school. 'And of course I did French A-level. Don't ask me to rattle away in it, I can't do it cold . . . but when I'm there, actually there with them, I can manage quite well.'

The singer had left the stage and couples were filling the tiny dance-floor. 'Same again?' Guy asked, rising to his feet.

'Not for me, thank you,' Jenny said firmly but Barbara smiled sweetly and pushed forward her glass.

When he had gone to the bar she turned to Jenny. 'Listen, Jen . . . he can't ask one of us to dance while the other's here.'

'Do you want me to go?'

'No, not exactly . . . but you can see the position.'

'It's OK, I'm tired anyway. I'd like to get some sleep.'

'You can't go home, Jenny!' Barbara sounded horrified.

'Why not?'

'Because Aunt Bea would go mad if she thought I was here on my own.'

'Well, where else can I go?'

Barbara was fishing in her bag. 'Here's the car key – sit in there. You can use the radio. And I won't be late, not really late.'

'What time does the dancing finish?'

Barbara shifted slightly. 'I don't know. Midnight, probably. But Guy has a room here so . . .'

'You wouldn't go up with him! Ba, you hardly know him!'

'Oh, Jenny, don't be paranoid. Anyone can see he's . . . well . . .'

'Posh?'

'You know I don't mean that. Anyway I'm not arguing. He'll be back in a moment. Are you going, or aren't you?'

'Give me the keys.' Jenny stood up as Guy came back to the table. 'I'm going now, Guy – I'm meeting a friend. I'll see you later, probably.'

He made polite sounds of dissent but even as Jenny threaded her way towards the door he and Barbara were moving on to the dance-floor.

Outside it was still daylight but a cool breeze had blown up and Jenny shivered in her sleeveless dress. She moved towards the car-park, trying not to meet the eye of all the unattached men who seemed to throng the Jersey streets.

It was a relief to be safe in the car, the door closed and locked, radio playing softly. She was still surprised by today's events. They had known Guy for such a short time, and yet prim Barbara was quite prepared to go to his room.

'I'm too stuffy,' Jenny told herself, and then she remembered Keir and felt ashamed – who was she to criticize? Except that she had known and hankered after Keir for three years. She remembered all Barbara's snide comments about Elaine being 'loose' – but you couldn't get much 'looser' than the first date.

'She won't do it,' Jenny thought, snuggling down in the seat. She watched the sky pale and then grow dark, saw cloud patterns disappear and reappear, and the lights spring up around the harbour. She dozed at last, waking to find Barbara rapping urgently on the car window.

'Open up, Jenny. Hurry! I don't want Guy to see you're here. He's just in the doorway, watching to see I'm OK. I told him to stay there. Oh Jenny, he's wonderful!' She flopped into the seat, her bracelets jingling as though she was trembling with excitement. 'He is so wonderful, Jenny, such a gentleman . . . and lovely manners. And his father's something very high up in Marks and Spencer.' She let out her breath in a theatrical sigh. 'I can't believe this is happening. It's too marvellous to be true.'

They were nosing out of the car-park, turning in the direction of the guest-house.

'Don't let on to Aunt Bea that we weren't together all the time. And don't mention Guy . . . let me do it in my own time. And don't wonder whether or not we did it, Jen. Because we didn't . . . well, not all the way, anyway.'

'I feel better, somehow, knowing you're in the same boat,' Elaine said. She and Kath were sitting side by side at the back of the booth while Dave and Alan stood at the bar, getting drinks.

'I wish I was as far on as you,' Kath replied. 'I'm still at the queasy stage. Every morning I lie there, cursing Dave, cursing the baby . . . it's the only thing that seems to help, a good swear.' She was about to say that the thought of her mother waiting downstairs didn't make things easier, but somehow she would have felt guilty telling Elaine about her mother's attitude. She had mentioned her mother's displeasure, but not the full extent of it.

'I tried cream crackers,' Elaine said. 'Nibbling them slowly, nearly all the time. That helps a bit.'

'You *are* getting all your vitamins an' everything?' Kath asked, alarmed. Elaine looked as though she lived on a water-biscuit diet. Or perhaps it was the swollen belly that made her arms and legs look like sticks.

'I eat loads of fruit. And fish, sometimes, and plenty of milk.'

'Are you OK for cash?'

'Yeah, thanks to Euan. He sends me a cheque every month. I wasn't going to spend it, but Alan said I should, and so did Jenny. They both said he'd be hurt . . . on account of Jed. So I use it, but I'm keeping track. I'll pay him back as soon as I'm working again.'

Kath smiled and nodded, wondering whether or not she should tell Elaine that she knew the baby *wasn't* Jed's. But the boys were coming back to the table and it was too late. Anyway, Kath decided, leaning back into the comforting circle of Dave's arm, better let Elaine think only Jenny knew. Otherwise she might think Jenny had told everyone, and that would never do. Obviously the father, whoever he was, wasn't going to be like Dave and do the decent thing.

As for Euan, tipping up in the belief that it was for Jed, that was fair enough – at least it would be keeping him happy. Poor Euan! They had never talked about it, not really, but she had a fairly clear picture of Euan's life in the posh house in Northumberland, with more rooms than you could count, according to Jed. He hadn't fitted in there, that much was plain. So he had clung to Jed, to all of them really, because they had accepted him for what he was and not tried to change him.

Suddenly Kath was intrigued by a new thought. What if Elaine wound up with Euan? Or Alan, for that matter? Even Keir? No, not Keir, she thought – pretty though she was, Elaine was not a high enough flyer for Keir. He had got what he wanted now, a girl with a souped-up car. Kath had seen it earlier on, roaring away from the Lockyers, Keir in the passenger seat, his arm extended behind the girl as she drove.

'Right,' Alan said, returning to the table to put down Elaine's drink and raise his own. 'To the holiday-makers. May Jenny come back sane . . . which, living with Barbara, is by no means certain.'

Dave sank into his seat when the toast was over and Elaine leaned towards him.

'What about your wedding plans?'

'Don't ask me, it all goes in one ear and out the other. If I had the arranging of it it'd be Easington Register Office and then Scarborough – no fuss.'

'That's what I'd like, too, but you know the position,' Kath said. 'I'm already in the dog-house – if I sneak off to get married I'll be in more trouble.'

'Do you think your mother will come round?' Alan asked.

'Eventually. When she knows she can't change anything, I suppose. You've got to be fair to her – this isn't what she bargained for.'

'She knew you'd get married one day,' Elaine said.

141

'Yes, but . . .' Kath bit back the words 'not to a miner'. Not even someone as well-balanced as Dave could take that kind of thing. Nor could she tell Elaine of her mother's ambitions for her to work and travel and see the world. She wanted to defend her mother, but she couldn't hurt Dave.

'She'll feel better when we get a house,' Dave said, wiping foam from his upper lip.

'A house?' Kath was taken by surprise. They had never talked beyond the wedding and the baby's birth.

'Aye, I thought we'd start looking soon, so that it's all tied up in time.'

'Can we afford it?' Kath hoped she sounded suitably impressed because she was.

'I think so,' Dave said complacently.

'There you are, Kath,' Alan said. 'You've fallen on your feet.'

'Where shall we look?' In her mind's eye Kath saw the hill above Sunderland, the new-built house snuggling against the slope.

'In Belgate,' Dave was saying, somewhat surprised. 'Where else would we live?'

Kath raised her glass, gloating at the thought of a house, a place of her own, the first place where she could do exactly what she chose. But even as she luxuriated in the very idea, a little part of her balked at his words: *In Belgate, of course. Where else would we live?* Was she condemned to live there for the rest of her life? Was that what her mother had meant about there being no escape – not for the wife of a miner, anyway?

13

2 August 1980

Jenny woke early, turning on her side to look at the clock, counting the hours until she would see Jersey dwindle to a dot and then disappear as the plane flew back towards England.

Since Barbara had become welded to Guy, Jenny had spent a lot of time alone, enjoying every minute, seeing far more of Jersey than she would ever have seen if Barbara had still been her constant companion.

In the next bed Barbara stirred. 'What time is it?

'Half-past seven.'

'Oh God, Jenny, I can't believe we're going . . . we've hardly got here.'

The voice was half-tearful and Jenny sought desperately for words of comfort. 'You can come back one day.' Would she come back to Jersey herself to see bays like picture-postcards come to life, to hear place names like Grève de Lecq and Corbière ripple on the tongue? Probably, but now all she wanted to do was go home.

She had showered, and eaten croissants and apricot preserve, and was going back upstairs to finish her packing when she heard Barbara weeping down the phone.

'Oh, daddy, I can't come home. Not yet. Tell mum, explain . . . he's so nice, daddy, you'd like him. His family live in Brighton, three cars . . . oh mum, has daddy explained . . . oh, you're a darling . . . I love you, I love you . . . yes, I've spoken to Aunty Bea . . .'

'So you're not coming back with me?' Jenny said, and when Barbara chattered out her explanations she took her friend in her arms and hugged her to show there were no hard feelings.

'This is my one chance to be happy,' Barbara said

143

solemnly, standing on the step as Jenny's bags went into the car. 'I'll write and I'll phone. Explain to everyone. Tell Kath it's turned out like I said: I *knew* it would all happen here. If I can stay on I can get him, Jenny, and he's exactly . . . *exactly* . . . what I want.'

'*So yours truly is really at odds with his kith and and kin,*' went Euan's letter. '*If I am found with an assegai through my cranium it will be the white folks what done it.*' Kath chuckled and turned the page.

The one thing I've learned in the past six months is that there's nothing much here for me, but then there wasn't much for me back in England so I might as well stay put here, for the time being. I can't imagine the old place without Jed. At least being here saved me from having to come to terms with all that, although I wished I could've been there with you all at the time.

Is Elaine all right? Not just financially but in spirit? I write regularly but it's difficult to know what to say. When the baby's born, though, I mean to make sure it knows about its father . . . terrible to call a baby 'it' but there's no alternative at the moment.

Kath drew her legs up under her on the bed and then remembered varicose veins and put them down again. It was uncomfortable spending most of her time up here, but preferable to the icy atmosphere below.

She looked at the clock: three hours before Dave would come up from his shift and collect her. She turned back to the letter thinking of Euan thousands of miles away, convinced that it was Jed's child coming into the world: when she wrote back she would have to watch her words.

Suddenly Kath felt her nose sting and then the tears start. She was always weeping now, getting sentimental or feeling compassion in the most unlikely circumstances. The nurse at the clinic had said it was hormonal, but she didn't know about Kath's trouble at home. 'I can't believe it's happening,' she thought. 'Not *my* mam.' A year ago she'd have bet money on her mother's infinite tolerance: only a few months back she had been so understanding about Elaine. So much for confidence that you knew everything!

Anyway, there was no point in dwelling on it. Her mother hadn't given an inch since she got the news, so she wasn't likely to do so now. And Jenny and Ba-ba black sheep were coming home today, so at least she'd have someone to talk to. Kath blew her nose and picked up the letter again.

Steven patted the folded letter in his back pocket to make sure it was still there. Not that he needed to keep it. He knew every word by heart, especially: *'You really should get the phone in, dad. It's the only way to keep in touch.'* There were more important things on his mind than getting the phone in, more important by a long chalk.

Now that the dratted Olympics were finished . . . or would be by tonight . . . there might be something else on the telly. A bit entertainment. Something else but a bunch of nancy boys prancing round in Persil outfits as though winning races had ever altered anything.

Sensing the dog was lagging behind, he turned to encourage it. 'Howay, lad. Keep up!' He tried not to see the whitened jowls, the yellowed eyes. Rheumy, they had called them in the old days. Rheumy! They didn't make good words like that nowadays.

He sat down on a low wall and brought the dog to his knee. There was no one around and he fondled its ears. 'Who's a good lad, then? Who's a bloody little petal of a dog, then? I wouldn't part with thou, bonny lad, for a king's ransom. We stick together, thee and me. Bloody telephone, indeed.' He got to his feet and went on his way towards the sea, remembering each line of Keir's letter with its tales of wining and dining and running about like a bull with its arse on fire.

'And for what?' he asked the dog. 'For what? To get his name on one of those lists they ruin programmes with. Run them off on lavatory rolls, most likely, there's that many of them on there.'

*

When Jenny woke she lay still, drinking in the familiar details of her own room. It was so *good* to be back. Utterly, utterly lovely! She raised her wrist and tilted it towards the window to catch the afternoon light that seeped through the drawn curtains. Five o'clock. She had been asleep for three hours, ever since she had arrived home from the airport in her father's car. 'Go straight up and have a lie down,' her mother had said, and Jenny had argued that she had only come from Jersey, not Australia. All the same, it had been bliss to mount the stairs and climb between cool sheets in a darkened bedroom that was unmistakably home.

There was a tap on the door and then her mother was peeping round, advancing into the room with a tray.

'I thought I heard you. Here's some tea.' She put it on Jenny's knee and sat down on the edge of the bed. 'It is good to have you back, darling. I know it's not for long, but I mean to enjoy the next week or so. I've heaps to tell you. Kathleen's been on the phone a couple of times and there are some letters . . .' She hesitated and Jenny looked up, sensing something wrong.

'I'm afraid Barbara's parents are coming round to see you. They rang just after you came upstairs, and I put them off for a while but they'll be here any time now . . . they're not the kind of people you can deter, are they?'

'Are they upset? They knew she wasn't coming back, I heard her telling them . . .'

'Oh, they know she's staying there, but I think they want first-hand news. There's a boy, isn't there?'

'Yes,' said Jenny 'but I hardly know him. I hope they're not expecting a blow-by-blow account. I only saw him once or twice.'

'Goodness gracious,' her mother said. 'No wonder they're worried, and what was her aunt thinking about? I'm glad it wasn't you . . . still, you'd have more sense.'

Jenny felt the blush mounting her cheeks. If her mother only knew . . . 'Any chance of a lift after tea?' she said hastily. 'I want to see how Elaine is.'

'Dad'll take you anywhere, I'm sure, he's so glad to have you back.' There was the sound of a doorbell. 'That'll be

146

them. You drink your tea in peace and I'll keep them talking.'

But Jenny couldn't settle to her tea, thinking of the anxiety below. She went downstairs after a few hasty sips to find Barbara's parents installed in the lounge.

'Well, now,' Barbara's father said, 'here's a pretty kettle of fish and no mistake!' But he didn't sound at all put out; rather the contrary.

'Is he nice?' Mrs Finch asked. 'This Guy, I mean?'

'He seems very nice,' Jenny said. What could she add? 'I think his father works for Marks and Spencer.'

'Not works,' Mrs Finch said. 'He's an executive there.' Jenny saw a gleam of amusement in her father's eye and looked hastily away. What else could she say? 'Barbara says she's going to get a job in Jersey. She really loves it there,' she went on desperately.

'That's all fixed,' Mrs Finch said complacently. 'She's going to help her Aunt Bea – while the summer rush is on anyway. Oh, I know you may say it's a waste of her education, but what's that compared with happiness?'

'Will you have some tea?' Jenny's mother said.

'No, thank you.' That was Mr Finch. 'We're going to a little do tonight. Masonic.' He patted his abdomen. 'Got to leave space for the fleshpots, haven't you? Thanks all the same.'

'What does Guy look like?' Mrs Finch said. 'Barbara says there's photos coming but I can hardly wait.'

'He's fair,' Jenny said 'And very tall. Lovely manners.' A wicked impulse to say he was quick to take girls to his room possessed her, but the pleasure on Mrs Finch's face and her own guilty conscience restrained her.

'Oh, well,' Mrs Finch said complacently, 'as long as he's all right with you, Jenny.' She turned to Jenny's parents. 'We value Jenny's opinion, you know. Such a sensible girl. And friends with our Barbara since goodness knows when.' She turned back to Jenny. 'You mustn't lose touch. Of course, Barbara may be home on the next plane if it doesn't work out. The fact that he's a lawyer . . .' She said the word with emphasis and then repeated it. '. . . lawyer . . . wouldn't weigh with us where Barbara's happiness was concerned. Still, if it does come to something you mustn't

147

lose touch.' She laughed, or rather simpered. 'She'll need a bridesmaid.'

When they were gone Jenny's father said something that took her breath away. 'I always thought that girl had no sense,' he said. 'Now I know where she got it from . . . or rather, where she didn't.'

'What would you have done if *I* hadn't come back?'

'Come and got you,' he said. 'First plane.'

'I should think so,' her mother said. 'Still, you did come back. Now, what do you want to eat and then where do you want to go? You'll be dying to get all the gossip.'

An hour later Jenny was mounting the stairs to Elaine's flat, shrinking from the claustrophobic lift.

'Oh God, Jen, I'm glad to see you back!' Elaine had changed shape since Jenny had last seen her. The gentle mound of her belly had become a sharply defined mass from which her arms and legs protruded, looking unbelievably stick-like.

Her face must have given her away for Elaine laughed. 'I'm massive, aren't I? Alan swears it's twins.'

There was an indulgent note in her voice when she spoke of Alan and Jenny smiled. 'Ah ha, so he's been dancing attendance, has he?'

'Oh no, Jenny,' Elaine sounded shocked. 'It's not like that. I couldn't keep up with Alan, he's far too brainy. He's been marvellous, though, while you've been away. Round nearly every day, wanting to know if I was OK, if I wanted to go for a drink, could he do a message? I'll never be able to pay him back . . . or you. And Kath's been through from Belgate heaps of times.' She smiled. 'We've been comparing our morning sickness. I'm the winner by a mile, so far.' Her face clouded. 'She's not getting on with her mam though. She doesn't say much but you can tell she's choked about it.'

'*Still?*' Jenny said. 'That's awful!' But even as she spoke she knew that her own parents, loving though they were, would take a long time to forgive a pregnancy. And remembering how lightly she had risked it she felt her mouth go dry with fright.

Jenny took a deep breath. 'Have you thought again about

telling Keir? I'm *sure* he'd help . . . and perhaps he has a right to know?'

'Oh no,' Elaine said, her small features suddenly animated. 'He has no right, no right whatsoever, Jenny. It was a moment's fun to him, nothing more. He was nice to me, though. Not pushy – gentle. That's why I wasn't careful like usual . . . because of what he said. You'll be all right with me, that sort of thing. You know what I mean?'

'Yes,' Jenny said carefully. 'I know what you mean.'

'Anyway,' Elaine said. 'I did it and I'm paying for it and it's nothing to do with him. He's gone off to London, anyway, with some girl he met when you went to that telly thing in Newcastle.'

'Really?' Jenny said, trying to hold her mug lightly, not grip it until her fingers ached. 'Toni? Was that her name?'

'I think so. Alan knows, he'll tell you. Anyway, just promise me you'll never let on to *anyone* and then let's talk about something cheerful. I'm dying to hear about Jersey.'

Jenny seized at the escape. 'Well, it's a good job you're sitting down. Barbara hasn't come back! She's staying on there.'

'Oh dear,' Elaine said politely, 'I will miss her, I really, really will.' They giggled then and Jenny felt some of the tension leave her. She would think about Keir tomorrow, when she could bear it.

She talked of Jersey, of the flowers, and the churches, the wide beaches, the narrow lanes, the perfume shops and tiny bays. She did not tell Elaine about the German hospital and she hoped she would forget it herself in time.

All around Jersey there had been signs of the German Occupation but this was different. Here, men had died as they dug, and had been buried in the walls with German efficiency. '*Krankensaal*' said a sign, and in the distance she had heard German marching music and voices, guttural and terrifying. She should have got out then but something had drawn her on, past the officers' quarters with their sprawling wax figures, an operating theatre, a ward with gingham sheets – and all the while the swastika and printed details of death and destruction. She had hurried on towards the exit to be suddenly confronted by a gift shop selling brightly coloured artefacts. 'They are feasting on corpses,' she had thought.

'Why did I go there?' she thought now. 'What did I want to see?'

'It sounds lovely,' Elaine said, thinking of flowers and the sea, and then the doorbell was ringing and Alan was there, smiling at the sight of Jenny, suitably astonished at the news of Barbara.

'Good old Babs,' he said. 'I didn't think she had it in her.'

They spurned Elaine's offer of more coffee and persuaded her to sit still and talk while the light faded outside the window and the last bird flew homeward.

'I've got to go now,' Jenny said at last, wondering if perhaps they wanted to be alone. It would be a fairy-tale ending if Alan and Elaine fell in love – and Elaine was in need of a fairy-tale of some sort.

But Alan elected to leave too. 'I just popped in to see everything was OK.'

'Bye-bye,' Elaine said at the door, seizing Jenny in thin arms. 'You don't know how nice it is to know you're back. I swore I'd cross me legs till you got here, Jenny.'

As they went towards the stairs they heard the bolts rattle home on her door. 'Poor Elaine,' Alan said quietly.

Outside in the street, the tower blocks loomed above them, and shop windows on either side were bright with colour. They walked across the town centre, almost deserted now, even the hum of the traffic muted. Jenny paused at the corner where their ways diverged, but Alan shook his head. 'I'll walk you home, Jenny. I'm not in the mood for bed just yet.'

'Doesn't your mother worry?'

He laughed softly. 'My mother, Jenny, will have been in bed an hour or more by now, so nothing's spoiling.'

They talked easily of their respective futures, of Elaine and Euan and Jed and Barbara. And Kath, who had more courage than most to put her degree aside and opt for domesticity.

'Will you miss Liverpool and university life?' Alan asked.

'Yes,' Jenny said. 'Yes, I will. I liked Liverpool, it's not a bit like they say. My room was home from home. And *of course* I'll miss university life.'

'I'll miss some things about Oxford . . . the friendship, the wonderful buildings, the occasional booze-ups . . . but on the whole I'm glad to be going. I want to get started. Sometimes I feel I've been toiling up this particular hill for a lifetime. And it's time my mother was free of me – more than time.'

Something quite desperate sounded in his voice but Jenny daren't risk meeting his eye. There was a quality in Alan, a kind of cold dignity, that stopped you from going too far, even in the friendliest of moments.

He left her at the gate, tipping his hand to his head in farewell. 'Good to have you back, Jenny. Sleep well.'

But sleep was hard to come by and when it came there were dark dreams of water and tears and a watcher on the shore.

14

16 August 1980

Kath's wedding dress hung on the front of the wardrobe. They had shopped for two days, in Sunderland and Newcastle, before finding two dresses to hire, one in cream for Kath, the bride, the other in apricot for Jenny, the bridesmaid.

'I hope it'll do up,' Kath said. 'I seem to be swelling up like a pumpkin and we can't send it back ripped.' She lay back on her bed, her bare legs protruding from her shortie nightgown, the rollers Jenny had put in for her an hour ago still in place.

'You can't believe the day is really here, can you?' Jenny said, thinking how Kath had changed, her face rounding, her breasts full and pushing upwards under the flimsy nightdress.

'No,' Kath said. 'I can't believe any of it really. It's not what we intended, is it?'

'What did we intend?' Jenny asked. 'We never really talked about it much. We said things like we'd be rich and famous or never have to mug up a book again, but that's about all.'

'You were going to be a great journalist,' Kath said. 'That will come true. And Alan will definitely be a judge. Keir as an MP?' She put her head on one side, as though considering. 'I think not. He'll be rich, though; I'm prepared to put money on that.'

'You don't like Keir, do you?' Jenny said.

'I don't *dis*like him,' Kath said. 'I just don't rate him – I never have. We went right through Infants and Juniors together. I kicked him once.' She savoured the memory for a moment before continuing. 'I don't think Jed liked him, either. Not really liked. They knocked around together at the comprehensive because they were both clever and both out of touch with the others, but I don't think they were buddies, like you and I. It was Euan who was Jed's soul mate.'

'And yet they were only friends for a year or so. You're right, though – Keir's a loner, now that you mention it. So is Alan, but in a different way. I'm a bit scared of Alan, if I'm truthful.'

'Because he's quiet? "Watch the quiet ones," me mam used to say. "They're the swift ones, the ones who don't talk about it."'

Jenny saw two tears form in Kath's eyes and brim over on to her cheeks. 'Don't cry, Kath. She'll come round. Wait till she sees you all dressed up.'

The tears came faster. 'That's it, though, Jen. I can't help thinking of the way it should've been, the way I always thought it would be. Her up here, fussing and carrying on, getting excited about her own dress . . . enjoying herself. Instead of which she's down there, sitting like the crack of doom. I don't even know what she's wearing, do you know that? Black probably, the mood she's in.'

She dried her eyes on a corner of the sheet. 'But I don't regret it, Jenny. OK, if I could do it again I wouldn't get pregnant, but I'd still be marrying Dave. Today! Whether or not she liked it.'

A van's rattling engine sounded in the back street, and there came a rap at the door.

'That'll be the flowers,' Kath said and Jenny went down to collect them, past Mrs Botcherby, seemingly indifferent and still in a cross-over pinafore, past Kath's brothers looking scrubbed and uncomfortable in formal shirts, unbuttoned at the neck.

'Aren't they lovely?' she said, holding out the tray bearing two bouquets and a selection of sprays and buttonholes.

'Very nice,' Mrs Botcherby said.

'I expect this is yours,' Jenny said, holding out an orchid backed by fern. 'It's beautiful.'

'Aye,' Mrs Botcherby said. 'Put it on the sideboard.'

One of the boys came forward and took the tray from Jenny. 'You take the two bunches, I'll see to these.' His back was to his mother and he pulled a face and raised his eyes to heaven.

'Thanks,' Jenny said fervently and retreated to the bedroom, to find Kath in tears again.

'Did she like her spray?'

'She loved it,' Jenny said.

'Liar!' Kath was smiling through her tears but then her face stiffened. 'I've had just about enough of the silent treatment, Jen. If she wants to be a pain in the arse, she'll have to get on with it. I'm glad I'm getting married, you're glad, Dave's bloody glad. Who else matters?'

There was a tap at the bedroom door and then it opened slowly to reveal one of the Botcherby boys, a mug of tea in each hand.

'Me dad sent these up. He daresn't leave the war-zone, but he says to tell you he's sending a posse down to get Dave to the church.'

Jenny took the mugs and he fished in his back pocket, producing two Blue Ribands. 'Keep your pecker up, Kath. There's worse things happen at sea.'

Somehow the tea and biscuits became a feast. They ate and drank, sitting cross-legged on the bed, talking wedding talk, keeping a wary eye on the clock so that they could both be ready in time.

'Euan sent us twenty-five pounds,' Kath said. 'A cheque and a lovely letter. And Barbara's dad came with a glass vase – he said it was a celery jar. You should've seen Dave's face when I showed it to him. I'll put flowers in it . . . if Dave grows them.'

'It's funny to think Barbara won't be here for it,' Jenny said.

'It's bloody wonderful, if you ask me. This wedding's been got together on a shoestring, Jen, and I can just hear what she'd've said.' She pursed her lips and gave a passable imitation of her former flat-mate. 'It's not that I'm being critical, Jenny, it's just that one has standards . . . I ask you, pie and peas for a wedding breakfast.' Kath's face creased into laughter. 'Don't look so scared. We're not really having pies and peas.'

'I know, dafty. By the way, my mum's picking up Elaine and Alan to bring them to the church. Alan's getting very thick with Elaine . . . it would be nice if something came of it.'

Kath shook her head. 'It won't. He's standing in for the others, that's all. Keir could've helped but it never occurred to him.' She leaned forward. 'You're not still keen on him, are you? Because if you ever marry Keir Lockyer you'll be marrying beneath you. And now take that outraged expression off your face and get me into that frock.'

'How do I look?' Kath asked, when she was ready. The cream gown flowed around her, the circlet of orange blossom crowned her upswept hair.

'I would say you looked lovely, Kath, but it wouldn't be enough. You look . . . just . . . well, I've never seen you look so nice, so beautiful.'

'Steady on, Jen. I won't get me head through the door at this rate.'

'Shall I go and get your mam now?'

'I doubt she'd come . . . you can try. And tell me dad I'm ready. He'll be like a cat on hot bricks . . . and he hates a shirt collar.'

Down in the living-room Kath's mother sat upright in an easy chair, dressed in a navy suit and hat and a pink high-necked blouse. Opposite, Mr Botcherby sweated inside navy pinstripe, a cigarette held between thumb and first finger.

'Is she ready?' He turned at Jenny's nod and looked at his wife. 'Up you go then, Margaret.' There was no movement from the chair. 'Come on, Meg, your place is up there with your daughter. I mean it.'

Jenny felt like an intruder as the two glances locked, and then Mrs Botcherby was rising to her feet and moving to the stairs.

'Have you given her something . . . you know, something borrowed, something blue?'

His wife kept on walking. 'She's got her nana's garter, she's had it for years.'

Mr Botcherby was fishing in his pocket, taking out a small grubby wash-leather pouch. 'Give her this for her shoe, then. Her other grandma's half-sovereign. And tell her to take care of it.'

When Mrs Botcherby had passed from sight he turned to

Jenny. 'It's a bad business, this – mother and daughter at loggerheads. She's a strong character, the wife. By God, it doesn't do to cross her.' And then, as if he had said too much, 'I'll just nip out and see if the lads are there with the car.'

A moment later Mrs Botcherby was coming back down the stairs, turning now and then to check that Kath, who was following her, was not going to trip over her long skirts.

'I'll be glad when I'm out of this lot,' Kath said but the face under the head-dress was serene, the hand holding the roses and gypsophila was steady. She looked down at her bouquet. 'I want these to go to Jed's grave, Jenny. Will you or Alan take them? I'll throw it to you first, don't worry, but after that . . .'

'They should go to your nana's grave,' Mrs Botcherby said, and Jenny saw Kath's chin quiver.

'Why don't I take mine to Jed?' Jenny said hastily. 'From all of us?'

She was glad when Mr Botcherby came back to scoop a pile of copper and silver from the mantelpiece. 'There's a right crowd of bairns out there,' he said. 'They'll be lucky to get a copper each.'

Kath saw Jenny's puzzlement. 'It's a Belgate tradition, Jenny. Don't you have it in Sunderland? We throw small change to the bairns as we leave for the church . . . if you don't, it's bad luck.'

Mr Botcherby took up the tale. 'I've got Freddy Larder out there with his brushes – it's good luck to meet a sweep on your way to get wed. She's got a gold coin in her shoe, that's for prosperity. And can we all get cracking now, before this shirt cuts off my breath?'

Jenny rode in the first car with Mrs Botcherby, two of Jenny's brothers on the tip-up seats opposite and a third in the front seat beside the driver. They didn't talk on the journey, even the normally irrepressible Botcherby boys seeming awed. Jenny waited in the church porch but the others went ahead to take their seats at the front.

And then Kath was there, her face serious behind her veil.

'All right?' her father said as the music struck up and Jenny moved in behind.

'Is Dave here?' Kath asked.

'He bloody well better be,' her father replied, pushing his chin forward and up in an attempt to escape his collar.

'Good luck, Kath.'

'Ta, Jenny. Your turn next.'

They were moving down the aisle, all the faces in the pews alight with pleasure at the sight of the bride.

Steven stood in the doorway to the backyard, listening to the news on the radio behind him. The yard was sunlit, making even the green-painted cree look fresh, but behind him the news was sombre. Commentators were reporting on Jimmy Carter's endorsement as the US Presidential Democratic candidate. According to them, Teddy Kennedy had managed to support Carter with one hand and ruin his chances with the other. 'His conservative policies look pale by comparison with Senator Kennedy's more radical proposals,' a nasal voice intoned. American politics were a shit-house, Steven thought, and the idea cheered him. If a nation as big as that couldn't find decent candidates, what hope did anywhere else have?

He was about to move out into the sunshine when the announcer's words arrested him. 'Workers at the Lenin Shipyard in Gdansk are striking in support of a dismissed colleague. The Polish authorities have not yet reacted to the situation.' He strained his ears but there was nothing more. All the same, there was some decency left in the world if workers could still unite.

'Mind out of the way, Steven, man. I'll miss it if I don't get a move on.' Cissie was pushing past him, shrugging into her coat.

'Where's the fire?'

'I told you, Steven. I told you twice when you were getting your breakfast. I'm going to see Botcherby's lass leave for church. She was our Keir's friend, mind on. Besides I like a good wedding.' She turned in the yard and grinned back at him. 'If I've got to suffer why should anyone else get off? Watch that pan if you go out – it's a ham shank, and I don't want it stuck to the roof when I get back.'

Steven fed the birds a handful of corn and went back into the kitchen to switch off pan and radio. If he didn't get out quick he'd get an hour of mush about bridesmaids' dresses.

'Right lad,' he said to the dog when he had fastened his boots and found the lead. 'Let's off out of it for an hour.' He wasn't spending a precious day out of the pit listening to a lot of sentimental clap-trap and likely a good cry to finish up.

He was making for the seafront when he recognized the figure ahead of him.

'Hold on there, Sidney.' The man wore a light white mac in spite of the heat, and carried a briefcase. Steven eyed it. 'That one of your perks?'

'If you mean did I get it for use on council business, Steve, the answer's yes.'

'So it is a perk?'

'Aye, lad. If it makes thou happy it's a perk. How are you, anyway?'

'Bearing up. I got out of the house before our lass gets back from the wedding . . .'

'Botcherby's lass? Our Mary's gone down there with the girls. Daft, I call it. When you've seen one wedding you've seen them all, I told them; but nothing else would do.'

'Weddings are for women, I reckon. If it was left to us they'd die out.'

Sidney was chuckling. 'We'd have the groom's night out and call it a day, eh?'

'That's about it,' Steven said. He could see Sidney was in a good mood now but it didn't do to pounce too soon.

'She's a few months gone, according to Cissie.'

'So they say. Still, it's the modern way – there's not many goes down the aisle with their cellophane on, not now. Any sign of your lad settling down?'

Here it was, the opportunity! 'No sign of a lass, if that's what you mean. He got a good degree, though, the best. He's off down London at the moment – they want him in television, or something like that – but his heart's still here. He'll be back.'

'I haven't seen him at ward meetings for a bit?'

'No. He's had exams . . . and of course he's been away. But he's kept up his subs, I've seen to that.'

Sidney was nodding. 'There's never been any doubt about you, Steven. You'd've got a nomination any time you wanted one if you'd fancied the council.' He caught the gleam in Steven's eyes. 'Well, somebody's got to do it. A place doesn't run itself.'

'I know that . . . and the lad knows it. He'll be back before long. I've got hopes of him landing a party job. Learn the ropes from the inside.'

'Not a bad idea. Mind you, we've got some good young'uns coming up. Still, there's always room for one more.'

They parted at the corner, Sidney to the council offices, Steven to walk by the sea and try to subdue his excitement at the thought of his son's future.

'Thanks for the lift,' Alan said as they climbed out of the car.

'No trouble, mate,' the man at the wheel said. 'I'll wait for you.' Alan was about to demur when the man's eyes flicked to Elaine and back to Alan. The message was plain: you couldn't let a woman in the last stages of pregnancy wander about a cemetery. He was only another wedding guest but he had volunteered the lift as soon as he heard of the errand.

'Thanks, it's very good of you.' Alan took Elaine's elbow as he walked between Jenny and Elaine, leaving their fellow guest behind in his battered Anglia. 'OK, Elaine?' She seemed to be breathing hard and he felt a flicker of panic, but she smiled up at him suddenly.

'Yeah, I'm OK, Alan, as long as you know how to deliver a baby . . . No, I'm only joking. All that's wrong with me's wind – too much chicken and trifle.'

Jenny hung back for a second and then moved up on Elaine's other side. She had changed from the hired brides-maid's dress to a sleeveless grey cotton shift, and felt comfortable once more.

'It all went off well, didn't it?'

159

'It was lovely,' Elaine said, with such fervour that Alan looked at her, startled. 'I mean, you could see Dave and Kath love each other,' Elaine went on. 'I know her mam's put out but she'll come round. She'll have to when she sees how well it turns out.'

They had reached the graveside and stood at the foot of the neatly tilled rectangle. *'John Edward Dawson dearly beloved son of Peter and Lavinia. Died 10. 1. 80. "Until the morning breaks and the shadows flee away"*,' the stone said.

Jenny bent to put her bouquet in front of the headstone. 'With our love,' she murmured quietly and straightened up.

'It doesn't seem like eight months, does it?' Alan asked.

'More like ten years,' Elaine said in flat tones.

'In all the rest of our lives I don't suppose so much will ever happen again.' Jenny was shaking her head almost in disbelief as she spoke. 'Jed dead, Euan and Barbara gone, Kath married, Keir in London . . .'

'And me up the spout,' Elaine finished for her.

'Never mind,' Alan said as they turned away. 'We all have things to look forward to, also.'

Jenny was silent, thinking against her will of Keir, of what it would have been like to have him beside her in the pew when they had sung 'Love Divine, all loves excelling' and the magic of familiar words like 'love' and 'obey' and 'worship' had swept over them. But he had not been there . . . only a telegram: *'Never thought you'd do it, Kath. Have a real Belgate do. Dave's a lucky man. Keir Lockyer.'* Not Keir just plain Keir, but Keir Lockyer – as though he were at one remove, now, from all of them.

Elaine had halted and was clutching Alan's arm.

'I hate to tell you this . . . it's like a real bad telly play . . . but I think this is it.' Suddenly she gave a little howl and looked down in consternation at a trickle of water falling to make a tiny puddle at her feet.

'What is it?' Alan asked Jenny, his face a mixture of bewilderment and terror.

'I think it means her waters have broken,' Jenny answered.

'I think it means I should sit down,' Elaine said faintly and they half-carried her to a seat.

'It isn't due for another week,' Alan stated as though defying nature to alter its plans.

'What are we going to do?' This time Elaine sounded afraid. Suddenly Alan was bending to lift her legs on to the seat.

'You're staying here, with Jenny. I'm going to get that chap to drive right up here, and then we're taking you to hospital. OK?' A moment later he was half-way down the pathway, running easily, looking odd in the formal suit as he leaped the corner turf and headed for the gate.

'Don't worry, Jenny, I'm glad it's here now. I want to get it over.' Elaine tensed suddenly and closed her eyes. There seemed to be tiny blue veins on her eyelids and then Jenny saw she had balled her fists and a tiny spume of foam had appeared at the corner of her firmed lips.

'Oh God, Elaine, I don't know what to do! Where are they?' But Elaine's eyes were open again now and her face was suddenly relieved.

'It's OK, it's passed. I'm all right now.' And then, 'You haven't told anyone, Jenny . . . about Keir and me?'

'No, you know I haven't. And I won't. But I still think you should tell him.' Then, seeing Elaine wince again, 'But let's not talk about it now. Just hold on.'

The car was at the end of the path, slowly negotiating the narrow walkway, the driver's face a comic mask of terror.

'Here's a carry-on,' he said as they half-lifted Elaine into the back seat. 'I said I would wait on but I never bargained for this.' His eyes went up to the rear-view mirror and he grinned at Elaine through the glass. 'You can call it after me if it's a little lad: Barry Michael. Two nice names for you. And the wife's name's Evelyn Mary. By, mind, you come out to a wedding and you wind up delivering a bairn. The lads'll never believe this.'

He talked non-stop on the way to the hospital, and they listened, grateful to him for relieving them of the need to make conversation. A nurse whisked Elaine away in a chair as the driver made his farewells, holding out his hand to Alan. 'Good luck, mate. You're in for a long wait if my first was anything to go by.' Alan accepted the good wishes gravely and Jenny warmed to his easy acceptance of the father's role, almost as though he was protecting Elaine.

When the man had gone they sat in the empty waiting-room. The walls were white-painted and the doors and skirting boards a turquoise blue. 'Give blood', exhorted one poster, and another offered advice on preventing VD.

'It's quiet,' Jenny said.

'It's Saturday afternoon. It's always quiet in hospital at weekends – no regular clinics, you see.' Alan grinned. 'If I sound like an authority on the subject it's because I am. Every time I've had a rugger or soccer injury it's been at weekends.'

Suddenly the nurse was back. 'Mrs Gentry is in labour now.' She seemed to stress the courtesy title. 'Everything's proceeding satisfactorily.'

'How long?' Jenny asked.

'Ooh, that's a difficult one. A few hours probably. You can go in and sit with her.'

'You go, Jenny,' Alan said, and Jenny nodded.

'I will . . . but you go in first while I ring Kath. If I ring now, I'll catch them at the pub.'

To her relief Kath was still there, saying goodbye to the departing guests and seeing to the arrangements for the knees-up that would begin at seven p.m.

'Oh God, is she all right,' Kath cried when Jenny gave her the news. 'Well, tell her to get cracking, and that I'm thinking of her . . . we're both thinking of her.'

'I don't think I'll make the do tonight, Kath. I'm going to stay with Elaine, if they'll let me.'

'That's fine. Don't worry, you see to her.'

'Any break in the cold war?'

'Not a chink, but me dad's trying to fill her up with Black Velvets, so there's still hope.'

Jenny took over from Alan in the side ward, almost laughing at the relief on his face. 'Kath sends her love,' she told Elaine, who was looking more and more exhausted. 'She says get it over quick and get back to the knees-up.'

As she'd hoped the joke brought a smile on Elaine's face. 'Some hope. But you go, Jenny – there's no point in hanging on here. They'll only chuck you out anyway, in a minute.'

'I'll go when they put me out, not before. Now, what shall we talk about?'

'Nothing serious, nothing sad, nothing to make me laugh. I feel as though I might split if I laugh. Tell me some more about Jersey, I loved it when you talked about Jersey.'

So they sat, holding hands, talking about the winding highways and by-ways of Jersey, of the magic bays where trees stepped right down to the shore and little red-sailed boats bobbed at anchor. They talked of flowers, Elaine's face relaxing as Jenny reeled off the names – hydrangea, petunia, carnation, lily.

'I've never had a garden,' Elaine said. 'I'll have one, one day, and I'll go to Jersey and see the zoo and buy enough perfume to bath in it.' And then her fingers were biting into Jenny's arm, and she was panting suddenly, like a dog, drawing away, her eyes wild.

'I think we'll go through now, Mrs Gentry,' a nurse said, appearing from nowhere.

'Is it over?' Alan asked hopefully, when Jenny rejoined him in the waiting-room.

'Not yet but it can't be long now.'

In fact it was three hours, during which they paced the floor, peered from windows, drank lukewarm machine coffee, lobbied passing nurses for news, and talked sometimes of Jed.

'He'd have been a wonderful father,' Alan said. 'When we were talking once, he said the '80s would be a decade of health: that was his professional dream. Privately, he wanted a home and children. "The only miracle that never ends," he said. I didn't agree with him. It isn't that I don't like children, it's just that I didn't have a clear picture of what it would be like to have a kid. I still don't. Perhaps today will wake me up a bit.'

'It's a girl,' the nurse said from the doorway. 'Seven pounds five ounces, and what a pair of lungs!'

'That's Jed's daughter,' Alan said, grinning hugely. 'Gobby from the word go.'

'I *have* to tell him,' Jenny thought, 'I can't bear to deceive him, to watch him building up a myth.'

In the end, though, she didn't speak, not even when they were allowed to see mother and baby, serene and

163

cocooned, and Alan was touching the tuft of baby hair and making more comparisons with Jed.

'She's lovely, isn't she?' Elaine said. 'I want to call her after Jed's mam – her name's Lavinia, isn't it? And after you, Jenny, so she'll be . . . what is Jenny short for? Jennifer?'

'Jane,' Jenny said, her throat suddenly tight. 'I'm sorry, but it's Jane.'

'Jane Lavinia,' Alan said.

'Lavinia Jane,' Jenny said. 'And I bet we call her Vinny.'

Elaine's eyelids were drooping but she held out a hand to each in turn. 'I couldn't have managed without you. Thanks ever so much.'

They bent in turns to kiss her cheek. 'Sleep well,' Alan said. 'And take care of the sprog.'

'I will,' Elaine murmured and was instantly asleep.

'Come on,' Alan whispered to Jenny. 'I don't know about you but I need a drink.'

They made for a nearby pub and drank the baby's health.

'I've got to admit it was a good feeling,' Alan said, grinning sheepishly, 'seeing a new life come into the world. Well, not exactly seeing . . .'

'But you were there,' Jenny said. 'We both helped. I know it made a difference to Elaine.' She longed to tell him how the sight of Keir's child had affected her, how she had imagined herself in Elaine's place, holding the baby, Keir protective at her side. But she couldn't tell, now or ever.

They ordered a taxi home and shared the cost, feeling suddenly reckless with earning so near at hand.

'I'll see you before I go to Coatham, Alan?' Jenny asked when she climbed out.

'Of course. And I'll be in London, so we won't be so far away.'

She was moving towards her gate when he called after her. 'And there's always Midsummer Eve. On the hill. 1989.'

Jenny nodded and smiled, but in her heart she doubted that the promise they had made to one another would be kept.

'It's over,' she thought. 'If this was youth, I think it's gone.'

BOOK TWO

1982

15

5 April 1982

'Alan!' Jenny held out both hands to him as he rose from the restaurant table. 'We shouldn't have let such a long time go by. It must be two years.'

'Nearly.' She had wondered if he might kiss her cheek, had been ready to proffer it, but instead he held her at arm's length for a moment, smiling, and then waited as the waiter pulled out her chair.

Jenny had chosen the restaurant in Soho's Rupert Street, when he had agreed to eat Chinese, wanting to show off her newly acquired knowledge of London and exotic foods. But her elation was disappearing fast. She had looked forward to seeing Alan again, but the man opposite was almost a stranger. And suddenly she realized the problem. He *was* a man now where once he had been a boy. And she must have changed, too, and perhaps be intimidating him.

Around them the tables were filled with Chinese diners, eating furiously and talking more furiously still.

'It's all a long way from County Durham, isn't it?' she said, grinning, and it broke the ice.

'Tell me everything,' he said, when they'd ordered and were sipping fragrant tea from tiny *famille verte* bowls.

'Everything?' She widened her eyes in mock-horror.

'The printable bits, then. I saw Kath in Belgate once when I was home, and of course Euan's letters are a mine of information! He seems to delight in passing on tit-bits. What do you think of him calling Elaine's daughter L. J.? He seems very taken with her photographs.'

'She's lovely. Very lively, bright, totally uncontrollable . . .' Jenny knew what Alan was going to say before he said it.

167

'Just like her father!'

'Yes,' she said quickly and was glad that their first course arrived to interrupt.

They talked about jobs then. 'I'm making a living . . . just,' Alan said. 'There've been times since I was called to the Bar when I've cursed the fact that I ever did O-levels, let alone law.'

'Was it that bad?' Jenny asked.

'Worse. I had to find a place in Chambers. I was a pupil, then, and had a master . . . a competent barrister. I did his donkey work at first, accompanied him to Assizes and Quarter Sessions, to the London courts. Gradually I began to get briefs while I was still a pupil, small cases, pleas in mitigation, that sort of thing. If I'd had to do one more petty theft that first year I'd have swung for my client. Then I started to get the odd County Court case and that livened things up a bit. But it was grim. I thought being a student taught you all about poverty . . . believe me, Jenny, I knew nothing. Still, let's hear about you and the BBC. It's marvellous that you've come to London. I felt we were losing touch when you were in Coatham and Norwich and God knows where. What are you doing now?'

'I'm a production assistant which is a grand title for a dogsbody. I do research, take care of nervous guests . . .'

'. . . and this is in radio?'

'Yes, I'm with Religious Broadcasting. It's . . . different. Interesting. I see a lot of Elaine – in fact I'm going round there tonight. You know she's living in Haringey?' Alan nodded as she went on. 'She works in a pub there, as a sort of singing waitress. She's really not bad, Alan, you must come and hear her.'

'What happens to the child?'

'Elaine shares a flat with two other girls, and if they're in they baby-sit. If not, she pays someone. It's not ideal but it seems to work.'

'Have you ever run into Keir while you were working?'

'No. He's not with the Beeb, he's the other side – and he's in telly, of course. Radio doesn't fraternize with TV. He's still living with Toni . . . remember Toni? . . . or so Kath tells me. And she's pregnant again. Kath, I mean. Did you know?'

They were interrupted by the arrival of a stream of steaming dishes, placed on little hotplates in the centre of the table.

'Yes,' Alan said when the waiter withdrew. 'It's strange – I never thought Kath would be the one to become domesticated so soon, but she seems to thrive on it.'

'It's a pity her mother doesn't agree with you,' Jenny said.

'I gather they're still at loggerheads?'

'Yes,' Jenny said. 'And Elaine's had a bad time with her family, too . . . at least, they've been no help to her. We're lucky, you and I.' His face clouded and she couldn't help a comment. 'Your mother's all right, isn't she?'

'She's healthy, if that's what you mean. But she's not happy.' He put his napkin to his mouth to mop an imaginary blob. 'I go up to see her as often as I can, but she's never been one for heart-to-heart talks. I suppose she's OK.'

Jenny saw his brow wrinkle and knew he was seeking a way to change the subject. 'I really enjoyed my time as a journalist in Coatham,' she said. 'I learned a lot, and it was the editor there who pushed me into applying to the Beeb. So I'm doubly grateful to him.'

'You're right in the centre of things now. Do you work at Broadcasting House?'

'Yes,' Jenny said. 'The hub of the universe, or so we like to think.'

'What's your opinion of this Falkland business?' The task force had sailed that day, HMS *Hermes* and *Invincible*, both aircraft carriers crammed with Harrier jump jets, leaving Portsmouth to the hooting of sirens and the cheering of crowds.

'It's a long way to go to fight a war,' Jenny said. 'Eight thousand miles. I can't really believe it.'

Alan nodded. 'It's ironic, really. The same men who've prepared that huge fleet of ships have received their redundancy notices this week, at least some of them have. All part of Mrs Thatcher's prudent housekeeping.'

'You sound quite bitter.'

'I suppose I am. I don't like the way the country is run at

169

the moment, though I don't see anyone who could do any better.'

'Have you ever thought of going into politics?'

'Not until lately. That was going to be Keir's province, wasn't it, before he got seduced by television?'

Jenny wanted him to go on talking about Keir and despised herself for it. 'Do you see him often?'

'About every four weeks. He seems to move about a lot, and I couldn't afford many nights on the town at the beginning. Keir has big ideas now, and he's taken to the jargon. Even I know some of it, just from hearing him talk. Cutaways, recces, vox pops . . . he sounds like Fellini at times. We must get together, all of us: you and Elaine and Keir and I. It's a pity Euan is still out in Africa. What about Barbara?'

'She's still in Jersey, and getting engaged soon, to Guy, who's a solicitor. The ring will be huge.'

'It would be,' Alan said gravely. 'It's taken them a long time to get around to it, though? Two years, almost.'

'I gather Guy is the cautious type. Barbara's over here at the moment, or she's coming over. My mum said something on the phone the other night, that there's been a message from her mum. You know what mothers are like.' It was the wrong thing to say, and she knew it. Alan's face had clouded once more and she had to do everything but handstands to persuade him to smile again.

May Denton had stayed on the shops side of the road to avoid a group of women talking at the crossing. Now she tried to negotiate the traffic further down, clutching her basket in front of her like a shield.

'Mrs Denton?' She pretended not to hear but the girl was not deterred. 'It *is* Mrs Denton, isn't it? Alan's mother?'

'Yes.'

'I'm Barbara Finch. I used to go to school with Alan . . . well, the same school but different years.'

May felt her arm being taken at the elbow. 'Are you going across? Good. My car's over there. Are you going home? I'll drop you . . . no, it's no trouble.'

170

In the end it was easier to let herself be carried along and folded into the big yellow car.

'I've come home to see my parents,' Barbara chattered on, as she started the car and pulled out into the traffic. 'I work in Jersey now. Have you been to Jersey? No? Well, you must go, Alan will probably take you. He's done well, hasn't he? You must be very proud of him.'

May nodded. The girl seemed to expect little else. She was very made up and dressed very stylishly, May thought, but her mouth never stopped.

'I must say it's a credit to you. The law's so expensive, isn't it? And you don't make any money at it, well, not for years, but *then* of course you do very well. Guy, that's my boyfriend, he's a solicitor but his brother's a barrister and it was a struggle even for them although his father practically runs Marks and Spencer. When Digby, that's Guy's brother, had his "call-night" in Middle Temple Hall he was actually called by a former Lord Chancellor.'

The girl was looking at May, expecting amazement. May tried to oblige but her heart had begun to thump uncomfortably.

'And of course there's rent of chambers and hotel bills and train fares and circuit fees and all those textbooks they have to buy because law's always changing . . . but of course you'll know about all that from Alan. And then the wigs and gowns. Guy's got one of Digby's black boxes lettered in gold . . . I've seen it, because Digby got a better one. But of course it costs a fortune, doesn't it? It's a rich man's profession, that's why I say you've been a marvellous mother . . . Is this your house? When you write, tell Alan to come to Jersey . . . Guy would love to meet him. And I'm sure Digby could be useful to Alan. I often tell him about the old days when we were young and daft.'

Long after May was in the house, door shut and bolted, she could hear the remorseless young voice: '*You don't make any money at it, not for years . . . rent of chambers, circuit fees, all those textbooks . . . gold lettering . . . gold lettering . . . gold lettering.*'

After a long while she got up and went into the kitchen. She filled the kettle from the measuring jug and spooned a measure of tea into the pot. She still had the house: there

171

would be enough money in the house to cover it all. She told herself this while she waited for the kettle to boil but her heart continued to thud until she took down the pills the doctor had given her and washed one down with water from the tap.

The news of Lord Carrington's resignation from the Foreign Office over the Falklands affair was balm to Steven. At least one of the bastards had had the grace to accept blame. He listened to the statement: 'I accept responsibility for a very great national humiliation.' To Steven's disgust Mrs Thatcher had refused to accept John Nott's offer to resign, which was a pity because Steven could see the man had shifty eyes.

'*She* isn't going, though,' he told Cissie but she was anxiously counting loops and couldn't have cared less. 'See that,' he told the dog. 'We're bloody well at war and your mam can't lift her brains above one plain, one purl. That's women for you.'

'He's right,' Cissie told the dog. 'We let the men talk about a crisis, we get on with doing summat about it.'

'Knitting socks for soldiers, then?' Steven asked, hoping to defeat her.

'No, Steven, I'm knitting a gansy for an old windbag.'

He was seeking for an answer when he heard the Prime Ministerial voice promising to reclaim the Falklands and dismissing thoughts of defeat by quoting Queen Victoria: 'Failure? The possibility does not exist.'

'Now there *is* a windbag for you,' he said triumphantly.

'It takes one to know one, Steven,' Cissie said and the bell clanged for the end of the round.

He took the dog along the allotments road, noting who was and who wasn't keeping their plot smart. If one fell vacant after he left the pit, he would take it. He could build a shed there for the days when Cissie was being clever.

For a moment or two he contemplated global conflict. If the Russians backed the Argies and Mrs T's fancy man, Ronald Reagan, waded in too, it could end anywhere. Cissie would be glad enough of him then, sitting under the stairs with the windows blacked out.

172

He thought back to that other war. He had been fourteen when it started and every day had brought fresh excitements. He had gone into the pit and worked like a madman to get coal to beat Hitler and hasten the new Jerusalem that would surely follow. And it had been good in those post-war years, Attlee puffing away on his pipe. A fly bugger, that one, but he got things done. And the Bevin/Bevan pair – different as chalk and cheese but good socialists. Nowadays they were bandwagon-jumpers, the lot of them. Steven kicked out at a tin and waited for the dog to chase it but it hung its head. 'What's wrong, marra? Come and tell your dad.' He sat down on a broken wall and waited until the dog came to him to have its ears fondled and its nose scratched.

'You know what, son?' The dog perked up, at least Steven could have sworn it did. 'Thou's worth a dozen of that London lot. Ponces, the lot of them.'

All the same, he thought, if Keir came back and settled down in Durham there could be possibilities. Definite possibilities.

Kath fished in the murky washing water for any item of clothing that might be lurking there and then set the tub to empty. A spasm of indigestion gripped her as she watched the churning water, and she reached in her shirt pocket for a peppermint.

If it wasn't for heartburn she would enjoy pregnancy, she thought, looking through the kitchen window to see if the pram was still. Brian was fourteen months old and could make the pram rock whenever he chose. Now, though, it was satisfactorily motionless and she switched on the kettle and spooned coffee into a mug.

She drank her coffee at the kitchen table, leafing through the morning paper for the second time, finding little solid news amidst the trivia and wishing they had enough money to buy a better, meatier newspaper. Perhaps they would call it a day after this baby, especially if it was a girl? Then, when the children were both at school, she could work and buy the things she wanted for the house. Get a better house, even, and banish the look of bitter satisfaction from her mother's eye.

'I don't regret it,' she told herself. How could she, with Brian as good as gold in his pram and a new baby starting to move? She burped gently, and scanned the headlines.

'*Argies Defiant*' said the banner headline. It was funny to think of Britain at war. Nothing would change in their lives: Dave would still go to the pit and play football on Saturdays, her brother too. And yet the lad down the street, Billy Davies, was steaming towards the South Atlantic on the *Invincible* and his mother was on valium.

The tub gurgled as the last of the water drained away and she rose to her feet to fill it again. Dave had promised her a better washer by the time the new baby's nappies arrived, but she didn't really care one way or another. She liked washing, didn't mind nappies even, especially when she was pegging them out in the garden, their fragrant damp folds blowing against her face. What she would like, what she missed, was a good talk – a real argy-bargy with someone, throwing around the pros and cons, acting devil's advocate to get the debate going. If only Jenny were here, or Jed. Even Barbara . . . she had always been able to get a rise out of Barbara. That was Dave's one drawback, he wouldn't be conned into an argument.

'I *am* in a low way,' she told herself as she showered Daz into the clean water. 'If I'd even welcome the sight of Barbara, I am on rock bottom.' To cheer herself up she took Euan's last letter down from the mantelpiece and sat down to read it again.

Dearest Kath, I think it's your turn for a letter but if not, why not! I hope you and Dave and the son and heir are thriving. Hopefully I'll be home to wet the new baby's head before the year is out. Mon Oncle keeps quoting the prices of air-fares home: is he trying to tell me something, I wonder? If he is I couldn't blame him as I'm the most god-awful farming apprentice that ever was. Apart from that, and to be serious, the poor man has more to worry about than an errant nephew. You've probably read about the Mugabe/Nkomo confrontation. Mugabe will have his one-party state before long, and after that I don't see a great future here for the likes of me. Besides which, I long to make music again. I've done one or two gigs at the club here but they're still into Coward and Novello so the applause was muted at best. Still I practise every day . . . the old paradiddles and triplets to keep the hands supple. I'll make Ronnie Scott's yet, see if I don't. Have you seen Elaine's L.J. recently? When I take my leave of Africa I will hire a mansion and summon you all to a reunion. Perhaps L.J. and your Brian will strike up a friendship and carry it into another generation? Give my love to Keir, if he ever deigns to visit Belgate. Jenny tells me he is something of a media-figure now. Jed always used to say his profile was the only worthwhile thing about him. Miaow!

Kath read the final affectionate lines and then tucked it back into the envelope. The water in the tub was almost hot and she was beginning to put in the clothes when she heard a piercing cry from the garden. Brian was awake! She smiled as she made for the back door, glad that at last she would have someone to talk to.

An abandoned fluffy panda lay on the grubby stairs, and Jenny retrieved it as she climbed towards Elaine's door.
'Jen . . . thank God you're here!'
As usual Elaine was in a flap, her face half made-up, L. J. under her arm, the baby mouth covered in chocolate. 'Can you take her a minute . . . keep her off your nice suit . . . I'll get a flannel. I'm going to be late if I'm not careful, and Bernie's not back yet. God knows what I'll do if she doesn't show up.'
'Calm down,' Jenny said, taking the soapy flannel and wiping the neat mouth. She liked the feel of the tiny figure on her knee, wanted to kiss the trusting, upturned face. 'There now, what a good girl you are. Let Auntie Jenny take the soap off.' She caught the towel Elaine threw in mid-air and mopped the baby's face.
'Stop panicking, Elaine. I can stay behind if Bernie doesn't get back, and then come on to the pub later.'

She dandled the child on her knee while Elaine made up her other eye, stretching her mouth wide as she did so.

'Did you see all those ships going off? I was crying. I mean, they won't all come back, will they?'

'I don't know,' Jenny said. 'We were talking about it at the office today. Most people seem to think it's a bluff and the Argentinians will back down before the task force gets there.'

'I hope you're right. Still, I can't make any better of it tonight, can I? D'you want a coffee?'

She made coffee at the sink in the corner, underneath a poster of Jeremy Irons and Meryl Streep in *The French Lieutenant's Woman*. Her bed occupied another corner, L. J.'s cot the third and the television the fourth. The wallpaper was peeling and there was a smell of decay. 'If only I could get her out of here,' Jenny thought, feeling the little girl brace her legs, wanting to get down.

'L.J.'ll be running all over soon,' she said tentatively.

'She already is,' Elaine said, sipping her coffee.

'What will you do when she needs more space?' Jenny persisted.

Elaine carried her cup back to the sink and swilled it clean. 'What I've been doing for the last two years, Jen. Cross the bridge when I come to it.' She put the cup on to the crowded drainer, balancing it carefully on top of a pile of other crockery. 'How's this boyfriend of yours? What's his name?'

'Bart, you mean? He's OK – or he was last night. No, he's better than OK, he is quite luscious, actually.'

'Is this *it*?' Elaine asked, smiling.

'Could be,' Jenny said, trying not to beam. 'But don't get carried away. We do not . . . repeat, not . . . have anything permanent going yet.'

'Don't take too long,' Elaine said. 'I could do with a wedding or a knees-up of some sort.'

'I had lunch with Alan today,' Jenny offered. 'He sends love to you both. Now that he's got your address he'll be round.'

'He doesn't have to,' Elaine said. 'He was good after Jed died and while I was still in Sunderland with the bairn, but I'm OK now.' She might have been talking of her dead

176

husband. 'She's forgotten about Keir,' Jenny thought. 'At least, she's forgotten he's L.J.'s father.'

'Alan knows you're OK,' she said aloud, 'but he still wants to come. He'd have contacted you before, but I honestly think he didn't have the price of a tube ticket until recently. And he wouldn't want to come empty-handed – which reminds me . . .' She fished in her bag and produced a brightly coloured ball.

'Thanks a bunch,' Elaine said as L.J. threw the ball in the direction of the crowded table. 'That'll be a big help!' Her words were sharp but her face softened. 'We can't stop here much longer, Jenny. You're right about that!'

The next moment Bernie was with them, her round black face perspiring. 'God, the tube, El. I thought I was there for the night. Hallo, Jenny.' She turned to L.J. 'And how's my best girl?'

They left the flat to the sound of L.J.'s giggles as Bernie pretended to pounce and then retreated.

'She's good with L.J., isn't she?' Jenny said as they went downstairs.

'She's bloody marvellous,' Elaine said. 'I couldn't manage without her and Carol.'

She paused when they got down to the street and checked in her bag. 'Got me key, purse, throat sweets . . . right. It'll only take two minutes to get there, that's the one good thing about the job. No, I don't mean that. He's good, the landlord. I've had a lot of rope off him when L.J. was sick and that sort of thing.'

'And he lets you sing?'

'Yes, as long as I get the glasses in on time. The lads . . . keyboards and bass guitar . . . they bring a crowd in. He likes the extra trade.'

Jenny settled in the corner, when she had been introduced to 'the lads' and the landlord, a lugubrious character whose face brightened when she ordered a brandy and dry ginger. She watched the regulars trickle in, young couples mostly, the occasional man on his own. They all seemed to drink lager, men and women alike, and they talked in subdued tones or sat silent, contemplating their glasses.

And then the guitar was being tuned, there was much checking of microphone and sound system and Elaine was singing, the hit song from Lloyd Webber's *Cats*. '*Midnight, not a sound from the pavement, has the moon lost its memory, am I weeping alone?*'

'She's really quite good,' Jenny thought, suddenly remembering that day in church a lifetime ago and Elaine's voice soaring through the hymn.

She caught Elaine's eye and signalled with an upturned thumb. 'Let her get out of that flat,' Jenny thought, 'that one room hell-hole. She can't bring Jed's baby up there . . .' Until she remembered that it wasn't Jed's baby, no matter how strenuously she and Elaine might wish it was. And suddenly the old pain was there, the longing for Keir and first love, which, no matter how hard you tried, was the one you could never forget.

16

12 April 1982

There was news of nothing except the Falklands on the radio. May listened for a while, sitting at the living-room window, her early-morning tea on a tray. She had been awake since daylight, glad to get up at last and get on with her task.

She had cleared the bedrooms one by one during the last week, destroying some things, boxing others and labelling them neatly. She meant to make it easy for Alan. Yesterday she had stayed in the front room all day, clearing, cleaning, washing down the paintwork and lightly starching the laundered curtains. She was more than half-way through the house if you counted by rooms, but the real work was still to come in the back room. Edward's desk, the bureau-bookcase, the corner cupboard with its box of papers, the two tin deed-boxes behind the settee: she must go through it all, page by page, in case there was anything there to cause distress. Edward had destroyed the letters when they arrived but there might still be something. After that there was only the kitchen and the back lobby, and perhaps the garden shed, and she would be done. She felt a little frisson of excitement at the thought and then a wave of fear at the mystery of what would come after. She carried the tray through to the kitchen and took down her pills from the window-sill.

While she waited for the calm they would bring, she thought of the girl. Barbara, that had been her name. Barbara, Alan's friend. Without her she would not have realized that it was far from over. Had Alan misled her deliberately or had she simply failed to grasp the facts? It didn't matter now, though, and she put it out of her mind. Instead, as the fluttering in her chest eased and her

179

shoulders relaxed, she remembered Edward's face when she had put his son into his arms. Until then he had seemed disinterested, even aloof, as though a baby had no place in his quiet home. He had taken refuge behind his paper while she sewed and knitted and read the books of names and christening etiquette and baby-lore.

And then he had looked on his son and everything had changed. She saw them in her mind's eye, in the garden under the apple tree, the baby lofted high in his father's arms or lying on the Otterburn rug while Edward dangled his watch . . . and then the picture that always came, the best of all – the tiny red-clad figure running ahead: 'Come on, mammy. Come on!'

'We were happy then,' she thought. 'We *were* happy then.' And it had been worth while. She had been the mistress of a lovely home, wife to a professional man . . . well, almost professional. He had always been so smart when he went to his office, a thorough gentleman. And now his son was a barrister, a man of law.

She sat on as sudden rain spattered the kitchen window, while the drug spread through her in a warm, comforting wave.

'He's made a bollocks of that,' Desmond, the producer, said wearily. He switched on the intercom to the studio: 'That was marvellous, wonderful. I think we might just do it one more time, though. I'll come through.'

He stood up when he had switched off and turned away from the glass which separated them from the bishop, sitting nervously behind the microphone. 'In God's name why did we book him, Jenny? I know it was my idea but that's no excuse. He hasn't had an original thought in fifty years. Still, I'll have to gee him up somehow. How much have we got?'

'Seventeen minutes forty,' Jenny said, checking her stopwatch.

'I'll persuade him to rejig his closing remarks and then we'll cut the whole thing to six minutes. Seven max.' He rolled his eyes to heaven. 'With a bit of luck no one will notice he's a moron.'

The next moment he was there in the studio and Jenny watched as he smiled benignly at the bishop, raised his hands in a flattering gesture and then leaned over the pages of the bishop's typescript to suggest amendments. She had been working in radio for a year now, so she was becoming used to the hypocrisy . . . the smiling congratulations that meant you had been unbelievably bad, the promises to set up another programme without delay while privately resolving never to use that contributor again. But here, in religious broadcasting, it seemed somehow worse.

There was a sound behind her and she turned to see Jeremy Elphinstone, the Reverend Jeremy Elphinstone, beloved of the audience for the Daily Service. 'There's a call for you, Jenny. A man. Do you want it putting through?'

'No, I'll take it outside. Desmond will be ages with the bishop.' The priest followed her out of the control room, the heavy cross around his neck swinging as he walked.

'Any grub?'

Sandwiches and vol-au-vents were laid out for the bishop, with a bottle of white wine and two cartons of fresh orange juice.

'Leave him something, Jeremy.' She had seen his locust-like onslaughts on hospitality before.

'Would I not?' he said, his expression pained as his fingers hovered over the plates, seeking the plumpest sandwich, the fullest vol-au-vent. As she lifted the phone, he picked up a carton of juice. 'I'll take this for the train, you won't need two. The bishop drinks plonk, anyway.'

Jeremy waited while the call came through. 'Bart? . . . yes, I am working. I've got a few minutes though.'

At the other end of the line he was telling her he loved her in tones so seductive that she felt the nape of her neck prickle.

'Yes,' she said, trying to sound businesslike for Jeremy Elphinstone, 'you've told me before. Yes, I should be home by five-thirty . . .' She looked through the glass to see bishop and producer still in conversation. 'Well, six at the latest. Yes, you know I do. Six o'clock then. Goodbye.'

'Is that your journalist?' Elphinstone asked, eyes greedy for gossip.

'Not *my* journalist, Jeremy. He's a friend, that's all.'

'Oh, would I suggest more, Jenny? Heaven forfend. Now, shall I go in and help Desmond with the bish so that you can get home to your friend?'

'I think Desmond can cope on his own, thank you.' She moved to the table and ostentatiously covered the remaining food.

'One more sandwich?' he pleaded, and she let him take the largest one. Tomorrow she would apply for a transfer. Even Farming Outlook would be better than this.

A map on the TV screen showed the 200-mile exclusion zone Britain had imposed around the Falklands. To Steven it seemed a damn sight too close to the Argentinian coast.

At his feet the dog stirred, moaning slightly. 'What's the matter, lad? Got a belly-ache?'

'He never touched his breakfast,' Cissie said from the doorway.

'He didn't make much of his food last night either,' Steven admitted. He lowered himself from the chair to the floor and took the dog's head between his hands. Close up the whiteness of its jowls was almost startling, the yellow of its eyes a reproach.

'He looks bad,' Cissie said apprehensively, twisting a tea-towel in her hands.

'I can see that for meself. Don't hit the panic button yet.' He stood up. 'Howay, lad. Let's us go down the beach.' The dog raised its head a few inches from the floor and let it subside.

'Vet!' Cissie said. She was already fishing in the back of her purse where the rent and electricity money was kept.

Steven reached for his coat and shrugged into it. 'I don't thinks he can walk, Cis.'

'Well, get a taxi. It's been a good dog to us.'

'I love you, lad,' he said as he carried the dog down the yard. There was no one to see or hear except the pigeons, and they had seen tears on his cheeks before.

*

Kath and Barbara met in the upstairs café of Dunn's department store.

'Kath! You do look well. And this is the big boy?'

'Barbara . . . lovely to see you. Yes, this is Brian.' Kath tried not to sound too doting but it wasn't easy.

'And Jenny says you're pregnant again? Well, I can see that. Sit down. I'll get a high chair.'

They ordered tea and cakes and a glass of orange for Brian.

'Now, tell me all your news. Is Dave OK?'

'He's fine. He's a deputy now . . . and he still plays football. He says he's getting past it, but that's just so I'll say he's not. What about you?'

Kath ate two cakes and drank three cups of tea while Barbara extolled the virtues of Jersey and Guy. She was tanned and glamorous, and for a second Kath felt a tinge of envy, until she saw the faint lines of what appeared to be discontent around Barbara's mouth. 'I'd've been in touch before, Kath, if you'd been on the phone. How do you manage without it?'

Inside Kath triumph blossomed. 'We are on the phone, Ba: 572963.' No need to say it had only been in a week and that she rationed herself to one call a day which mounted up to an alarming ninety-one a quarter.

Barbara copied the number into her neat pocket book.

'Still making lists?' Kath enquired sweetly.

'I have to . . . there's so much happening in Jersey. The social life is out of this world. If I didn't keep track of it on paper I don't know where I'd be. And it's not going to get easier.' She leaned forward and lowered her voice. 'I'm getting engaged in two weeks.' She twiddled the fingers of her left hand. 'We've got the ring, it's enormous. Diamond, of course, one and a half carats. We're waiting till my birthday – Guy's very traditional about things like that. I would have liked . . . still, we don't want to rush. Marriage is so important Guy says you should treat it with respect.' Her eyes flicked from the baby placidly chewing a rusk in the high chair to the mound of Kath's belly.

'Guy sounds a very sensible guy,' Kath said, glad she could laugh at the pun and ease her clenched jaw. 'Well, I hope you'll be as happy as Dave and me.'

'And what about Jenny and this editor?'

'He's a journalist, actually. I haven't met him but Jenny says he's nice.'

'Alan's doing well. I met his mother the other day. She's a little hard to get through to but very lady-like. I told her straight . . . I think you've been marvellous, I said, putting a son through an expensive profession like the Bar. I don't know how she did it. Did I tell you Guy's brother is a barrister?'

The baby had been good for a long while. Now he decided to change course. He reached for the orange juice which Kath thought she had put out of range and lifted it in two pudgy hands.

'Dink', he said cheerfully, and decanted it into Barbara's pleated lap.

Keir had made sure Toni would be home first. He loathed having to peel potatoes or delve in the freezer till his fingertips burned. He found her in the kitchen poking doubtfully at two pieces of fish.

'It's supposed to be monkfish . . . what do I do with it?'

They settled for frying it in hot oil and serving it liberally basted with bottled tartare sauce and surrounded by mixed vegetables from the freezer, boiled in a pan.

'What's it like?' Toni said anxiously as she sat down.

'Fairly vile . . . could be worse.' Keir had wanted to tell her about the trouble they had had getting decent visuals for his piece on the new Barbican Centre but she seemed obsessed with some tale of a PA who'd been fired for running up her expenses.

'And it's so unfair. *They* waste money all the time and *we're* criticized for spending farthings. I told Jack I'm going to go – I expected he'd at least argue with me, but he just shrugged. They're always trying to tell us we're expendable. "Twenty people queuing up for your job" – that's the message. I have had it up to here with Jack, I kid you not.'

'We did the Barbican piece today . . . twenty minutes. It should edit nicely.' Keir wanted her to tell him he'd done well. He still needed her approval although he realized now that he would go farther than Toni: he was more innovative and had a better feel for a picture. Toni was good but pedestrian.

He lined up his knife and fork, leaving a sizeable portion of his meal untouched, and tried to give her his undivided attention. She had stopped moaning about petty irritations and instead was talking about a new show Global Television was putting together. It was an independent company and that was where the future lay.

'They want to work with a studio audience, the same one each week, to get a real feel of what people are thinking about current topics. They've got this idea about having a triangular forum and the presenter in the middle, moving around.'

'Who's producing?'

'Annie Love's the exec. Don Philips is producing, but Annie will pull the strings. You met her, remember, when we had the end-of-series party? She went to Global just after that.'

'Was she the one who was obsessed with fitness?'

'That's right. She runs every morning on Primrose Hill. Her flat's near there. She's got legs like whipcord – horrible really. She's probably a dyke. However . . .' She looked at him and grinned. 'This show is going to be red-hot, if I'm any judge, and they need a good PA, so little Toni is going to offer herself for the vacancy.' She put her head on one side and leaned forward to look into his face. 'Well, should I? Am I up to it?'

'Yes, of course you are. All the same, I wouldn't rush. I know Jack's a pain, but at least you know your job's there for the foreseeable future. New ideas have a nasty habit of fizzling out. You could get caught between two stools.'

He excused himself from the table then, on the grounds of urgent revision of his notes for the next day, and went into their shared bedroom. His running shoes were where he'd thought they'd be, at the back of the shelf in the wardrobe. He put them into a sports bag along with the black tracksuit with the yellow trim, and head and wrist bands, still in their cellophane covers. He had bought them on a whim and then decided against using them. Now, though, they were going to come in nicely.

They came back to the silent house carrying the collar and the blanket they had wrapped the dog in for its last journey to the vet.

'I'll put the kettle on,' Cissie said but Steven's stomach churned at the thought of tea.

'We should have brought him back here,' Cissie said when the tea was brewed.

'Talk sense, woman. Where would we've put him . . . hacked up the yard?'

'There's room at the front. At least he'd've been in his own home ground.'

'What's the odds now, any road? He's dead. It makes no matter.'

They had agreed to a lethal injection when the vet had outlined the facts: a malignant growth that could not be treated. 'Will he suffer?' Cissie had asked and when the answer was yes they had both agreed to the vet's suggestion.

The dog had licked Cissie's hand as the needle went home and Steven had patted its flank and told it for the last time it was a bonny lad.

'Never mind,' Cissie had said as they left without him. 'He had two years of being loved every day. At least he had that.'

Now they sat either side of the fire, trying to ignore the collar discarded on the table, the bowl of water by the back door, the chewed ball in the hearth.

'Our Keir'll be upset when he hears about Chips,' Cissie said suddenly and Steven nodded agreement. Privately he wondered if Keir would even remember they'd had a dog, let alone mourn him. 'If we had the phone in we could ring and tell him,' Cissie went on

'Don't be daft, Cissie – ringing London to say a dog's dead! It was only a mongrel, you know, not a Russian Borzoi.'

Cissie put her cup down in its saucer so that they rattled together. 'Steven Lockyer, it's a good job I know what goes on in your mind because if I listened to some of the things you say I'd take a brush to you.'

She got up then and emptied the water-bowl, rinsing it under the tap and putting it to drain.

'I'll get rid of that by and by,' Steven said, ashamed to be sitting idle.

'We'll keep it by,' Cissie said and he didn't argue. If it comforted her to think they would have another dog one day, that was fine by him. But he knew there would be no other dog. They would never find the match of the lad and, besides, he was getting too old to start again. He was fifty-seven years old, and somehow that seemed only a clock's tick away from seventy. Threescore years and ten: the finish.

It wasn't death he feared. He had faced death more than once in the pit and never flackered. What he feared was giving up his job, no longer being a collier. After that would come the loss of his faculties, his strength. Most of all he feared fear itself, fear that might reduce him to a cringing wreck in front of a bunch of bairns in white aprons, brought in to spoon-feed him and change his nappies when he papped. He found he was sweating slightly, his heart thumping. It wasn't fair, that you came to this in the end. He had looked at Cissie once or twice lately, wondering if she too was afraid, but she went calmly on with her knitting, cackling at the telly, gossiping on the corner, as though she had all the time in the world.

If only Keir would come back. Would he do so when he got this television thing out of his system? Perhaps he should speak to Cissie about the phone: it would be one way of keeping up with the lad.

Bart brought a bottle of wine into the bedroom, rattling in a flower-vase of ice. 'For after,' he said and Jenny laughed. The way he spoke about making love, quite openly, had shocked her at first. Now it thrilled her.

Before, it had been something she could only contemplate in complete intimacy, close to someone, preferably in the dark, each of them moving stealthily towards a mutually desired end they never acknowledged. There had been two men since Keir, and the pattern had always been the same. And then she had met Bart at a press viewing, and at eleven thirty in the morning, in the middle of a gaggle of critics, he had said, 'I want to take you to bed.' She had not slept with him that night. It had taken two weeks of lunches and dinners and silly little gifts left at Reception addressed to La Giaconda, c/o Miss Jane Sissons, before she let him into her bed. Now, to all intents and purposes, they lived together.

If it hadn't been for Bart she might have invited Elaine to share her three rooms in Earl's Court. At least there was a garden, and only one flight of stairs. But Bart existed and at the moment she was too intoxicated with him to allow any other consideration to intrude.

They made love slowly, laughing sometimes, each reaching to move the hair tenderly from the other's eyes or stroke a particular plane of flesh.

'Good?' he said, when it was over.

'Yes,' she said, knowing he would want more than that.

'On a scale of ten, where?' he asked.

She pretended to ponder. 'Ooh, six . . . six and a half, somewhere like that.'

He reached for her then, pinning her arms, pressing his mouth on hers, entering her again even as she apologized and awarded him ten out of ten.

Afterwards they drank wine from misted glasses, resting the bases on their flat bellies, giggling as they let go and called 'No hands'.

'I needed that,' he said at last.

'What?' she asked, feigning ignorance.

'This wine,' he said, leaning over carefully to kiss her when she pouted disappointment.

'Where do we eat?' he said. 'I can just about afford haddock and chips twice.'

'You can't get decent fish and chips in London,' Jenny said loftily.

'I take exception to that,' he said, deepening his South London accent. 'Just because we favour plates and cutlery . . .'

He ducked the pillow she threw and then lay back laughing while she went for a shower and then to cook for him in her tiny kitchen.

17

19 April 1982

Keir had willed himself to wake without the aid of an alarm. He turned carefully in the bed so as not to disturb Toni and looked at the clock. Six-forty-five. Perfect! He eased himself from the bed and picked up his sports bag from behind the cheval mirror, where he had put it last night.

In the bathroom he sluiced his face in cold water and applied after-shave. There was just the right amount of stubble on his jaw – it mustn't look like a deliberate pick-up. He combed his hair and then retousled it before donning the sweatband. Too gimmicky? He regarded himself in the mirror, head on one side, and decided he looked better without.

He settled for draping a yellow towel scarf-like around his neck and then let himself quietly out of the flat. Annie Love started her run at seven-fifteen precisely. He wanted to be two minutes behind her and gradually overtake. It was essential that he looked sweaty and exerted. She was no fool, and if his ploy was to work she must be off her guard.

Keir felt fine as he drove towards Primrose Hill but when he had locked the car and started to run his knees ached and his breath wheezed painfully from his lungs. After a while, though, he found his stride and began to lope purposefully along the track he had seen Annie take the last two days, observing from the safety of his car or a convenient tree-trunk. Toni had woken the second day and had queried his absence. 'I'm sleeping badly,' he'd told her. 'And these light mornings tempt you out of bed.' In reality he ached and longed for his bed. If he brought this

189

off he would have a week of late rising and sod getting in on time.

As he came through the gate he expected to see the grimly determined navy-track-suited figure ahead, red hair caught back in a Batik scarf. There was no one except two middle-aged men jogging comfortably, side by side. Keir broke into a lope, seeing the netted peaks of Snowdon's aviary in nearby Regent's Park, hearing the traffic moving around the park's perimeter.

He kept on grimly until he had quartered the park and knew she was not there. Shit! The one morning he had geared up for and she wasn't out. Still, there was always tomorrow. He turned and slowed his pace. Back to the flat, shower, eat . . . he remembered Toni asleep in their warm bed. If she was still there . . . He quickened his step until he was back at the car, sweating partly with exertion and partly with anticipation.

Jenny had hoped Alan might look for her when he came into court, since she was there by his invitation, but his face was set and he looked nowhere except at his brief.

'It's only a plea in mitigation, Jen,' he had said when he invited her. 'The man's pleading guilty to his sixth robbery with violence, and I'm supposed to persuade the judge to go easy on him! All the same, it's Court No. 1 at the Bailey. If you can hang on, we can grab a bite somewhere when I'm finished.'

By mistake the cab had dropped her at the new courts. She had gone in through a narrow concrete passageway, marvelling that anywhere could be so bleak and unwelcoming. The concrete stairs were worn where she climbed, passing a girl crying and seeing two youths sitting with their backs to the wall, heads slumped on their chests in seeming despair. She came to an Enquiry window and asked directions. 'Wrong building, love. Go out and round to the other side, you want the old courts. Just ring the bell.'

Jenny walked round the building, taking in the ugliness of the mock stone, the mock-medieval absence of windows. She had to cross the road and crane to see the gold figure of Justice on the green dome, and even that was a disappointment.

Once she was settled in the public gallery of No. 1 Court she felt a greater sense of occasion. The walls were oak-panelled and a high dome seemed to swallow the voices of the participants. An attendant in a red-cuffed uniform had shown her to her seat. Now he sat impassively, guarding the door as though against intruders.

The coming and going in the well of court intrigued her, all the smiling and whispering as though everyone who worked there was having a ball. When Alan rose to his feet, though, she realized it was a serious business. His voice was cold and harsh and although he was asking for leniency he showed no pity for his client's shortcomings. Sometimes, as he spoke, he flicked at the tails of his wig with a pencil. Quite often he hitched up his gown on his shoulder as though getting a firmer grip on his facts.

This was a new and different Alan, face stern under the iron-grey wig. And yet she had always known there was steel in him, right from the beginning. Suddenly she saw him, standing on the river bank, face implacable while Jed struggled for survival. It was only a single frame, sprung up in her mind's eye and immediately extinguished, but it made her uneasy. Alan was prospering in his career; Jed's had been cut short

When she brought her attention back to the court the judge was speaking.

'Your counsel has been honest about your shortcomings. I commend him for that and it goes in your favour that he did so.' So Alan had been clever, after all. 'Nevertheless, this court cannot countenance violence, especially in the pursuit of gain. I sentence you to five years . . .'

When the court adjourned Jenny waited in the busy street until Alan appeared, unfamiliar in black jacket and striped trousers, already raising his hand to a cruising taxi.

'You were very good. Most impressive . . . even awe-inspiring.' Again the shutter opened in her mind, showing the implacable face, the roaring water, Jed crying out for help that never came.

'Thank you, Jenny, for the kind words but I'm still a complete novice. My knees were trembling, I promise you.

191

Still, there *is* something I'm rather proud of today.' He was taking a folded paper from his pocket and turning to a not far from back page.

'There, that's me. I couldn't use my own name, it's not the done thing, so I used my mother's maiden name, Murfield.'

He was pointing to a short article on constitutional law, based on the transfer of sovereignty from Britain to Canada which the Queen had signed at the weekend.

'I didn't know you were interested in this sort of thing.'

'I'm not.' He was grinning. 'I sat with a towel on my head to mug up that lot. I did it for the money. I've had the odd thing published over the last year, mostly moans about life at the Bar, one piece in *Punch* about northern versus southern humour. But this is my best effort so far, financially and literally. I went back to John George Lambton for that, the father of the Commonwealth.'

'Our Lambton . . . the Durham Lambtons?'

'The same. John George, first Earl of Durham, put forward the idea that men on the other side of the world could not always be governed like the Home Counties. We've got a lot to be proud of in our neck of the woods, Miss Sissons.'

They had steak and kidney pie and red wine in a raftered pub, and good coffee in satisfyingly large cups. It was easy then to relax and talk, and forget foolish fantasies. This was the old Alan, friendly and knowledgeable. She walked back to the Old Bailey with him and he hailed a cab for her.

'Life's going to be easier from now on,' he said, handing her in. 'Let's see more of one another . . . and all the others. I'll get in touch with Elaine this week.'

Jenny waved as the cab pulled away, and turned in her seat to watch him standing, one arm raised in farewell, until he was out of sight.

May was on the landing when she heard the doorbell. For a moment she contemplated staying there, out of sight, until the caller went away. But if she did that they might continue to ring, and then to knock or even to rattle the letterbox. It was easier to come downstairs and open the door on the chain

'Mrs Denton?' It was the woman from across the road. 'You haven't been out for a day or two . . . and I just wondered if you were poorly?'

'No. I'm all right.' She was about to close the door when she remembered the conventions. 'Thank you. It's kind of you to bother but I'm quite all right.'

The woman was not moving away and May felt her heart begin to pound. 'Is there anything you want at the shops? Or perhaps I could cook something for you? I've been making ox-tail soup, would you like some? It's awfully nourishing.'

It would have been easier to accept but May had cleared the kitchen. She had tea and coffee and biscuits left. That was all she needed.

'Thank you so much. It's very good of you to offer but I'm all right. Really.'

When the woman had retreated down the path May bolted the door and stood with her back against it for a moment. Her heartbeat had steadied but her legs were suddenly weak, so that she had to hold on to the stairs to make her way along the hall to the kitchen.

She laid out the silver polish and cloths and then collected the few small items of silver from the front room. Alan's christening mug, the mote spoon, the one candlestick she had kept when she sold all the other silver, and finally Edward, sepia-tinted inside an Art Deco frame.

She laid on the white liquid, watching it dry almost instantly, polishing and burnishing until everything gleamed.

Kath parked the pram outside her mother's door and pointed out a huge green articulated lorry to the baby. 'Look, Brian, lorry! It's going to the pit where Daddy works. Daddy!'

The baby smiled at the word and Kath bent to unfasten his harness. 'There now, let's go and see nana.'

Her mother was kneading bread dough on a floured surface, turning it, kneading, turning again.

'I brought the bairn round to see you.'

'So I see. There's tea on the hob.'

Kath held the baby on one arm and poured her tea with her free hand. She sat down in her father's chair, her foot on the fender, and watched as her mother heaved the mass of dough about as though it were feathers. After a while she lifted it into a huge earthenware bowl and covered it with a clean tea-towel, then carried it to the hearth and put it down inside the fender to prove.

'Anything fresh?' Kath asked.

'Not particularly.' It was the way her mother always spoke to her now, never giving an inch. If something tremendous had happened within the family her mother's answer would still be 'not particularly'. To say more would be fraternization and that was not allowed.

'I came round to tell you we've got the phone in at last. I've written the number down for you.'

'I'm not likely to ring you, am I? You're only a stone's throw away.'

Kath felt anger rise up in her but she quickly quelled it. If this was ever to be sorted out, and it must be for the sake of the bairns at least, she must not rise to the bait.

'Well, it's here anyway. I'll put it on the mantelpiece. Lads all right?'

Her mother nodded.

'And me dad?'

'He's his usual. That bairn's got a stye coming.'

'No, he hasn't. He's just been rubbing it.'

'Well, it looks like a stye to me. Please yourself.'

'I haven't got much longer to go now. It's gone over easy this time.'

Her mother's answer was to reach forward and poke the fire.

'Dave says we should book up for a holiday,' Kath continued. 'I don't know, though. It'll be the back end before we can get away.' She wanted to tell her mother how good Dave was, how well he was providing for her and Brian, how big his plans were . . . she had rehearsed it all the way up and it had sounded good. Now she knew it would be useless.

'What do you think about the war, then?' She was getting

194

desperate and the Falklands was all that was left. 'Dave reckons the Americans want it finished. He says they'll put pressure on Maggie.'

'She doesn't bend to pressure, that one.' Her mother's face showed a grim satisfaction. 'I don't give her much but I'll give her that. She's a woman. She's strong when she has to be.'

'Well, I hope it's over soon. There's bound to be some killing if it goes on. It's surprising it hasn't happened yet.' On her knee the baby had started to squirm. She picked him up and turned him to face her, lifting him high in the air. 'Are you fed up, pet? Want to go walkies? Well, give your nana a love, then, and we'll go.'

She held the baby out, hoping to see her mother's arms extend a welcome. But the big hands with their flour-rimmed nails stayed still on the arms of the chair, even when Kath extended her own to bring the baby's face close. Her mother held up her cheek momentarily to the tiny mouth and then bent down to twitch the tea-towel from the bowl. 'Well,' she said as she replaced it, 'I can't sit here as though I had corn growing.'

As Kath moved to the door her mother was already busy at the table, greasing and flouring the tins that would hold the loaves.

'See you,' Kath said, trying to keep her voice steady.

'Don't forget the gate as you go out,' her mother said shortly. 'I'm sick of dogs scavenging the bin.'

18

25 April 1982

It was Keir's sixth visit to the park and he was beginning to despair of ever meeting Annie Love this way. Life was a shit, he thought, as he began to jog down the path. She had been in this park every fucking day until he needed her to be there, and then she had stopped.

Time was running out for him. Toni was going to put in for the new programme and if she didn't get it someone else would. There was no point in his putting in for it in the normal way: he didn't have enough experience even to merit an interview, though he could do the job better than anyone if he was given the chance.

He rounded the corner, hoping to see the gaunt figure of Annie Love ahead of him but the metalled pathway was empty. He jogged on, reflecting on the last two years, mourning his lack of progress. He had got into television, that was something, but it would take years to get anywhere near the top if he had to suffer boards. The only short cut was to catch the eye of someone in power and play the sex card. Women had been doing it for years, why shouldn't men?

And then he saw her, sitting on a bench, feet apart, hands and head dangling loosely between her knees! As he watched she raised her head high and inhaled deeply before flopping down again. She was obviously doing some kind of breathing exercise, but he wasn't to know that. He put on a sprint, afraid she would raise her head again before he reached her.

'Are you OK?'

She lifted her head and Keir feigned surprise. 'Oh . . . hello. It's Annie Love, isn't it? Are you all right?'

She was grinning, showing perfect if wolfish teeth. 'I'm fine. Winded, but fine. Do I know you?'

'We met at the "Vision-On" end-of-series party. Keir Lockyer.' He held out his hand and Annie Love took it briefly.

'Well,' he said, 'if you're not in need of resuscitation, I'll be on my way. Unless you're ready to go again . . .?'

They jogged off side by side, Keir accommodating his stride to hers. When she had stood up her eyes had raked him from head to foot. Please God he'd passed muster.

'So you're in the business, too?' she said between breaths.

'Yes, I'm with Eastern. I joined them when I came down from Oxford. I'm doing specials at the moment but I'm hoping . . .' He hesitated just long enough. 'Well, I mustn't get on my hobby-horse.'

She was panting now, her lips drawn back from her teeth in a way he found oddly erotic.

'No, go on. Tell me.'

'Well, I'm hoping to get into current affairs. I have this thing about really getting to know how viewers are feeling . . . you know, get them in a studio, get them going.' Mustn't outline her project too precisely. It had to look totally coincidental.

Annie Love was increasing her pace now that the park gates were in sight. God, she wasn't going to bite.

'Give me a ring next week. I'm with Global now. We can talk. No promises.'

Jenny flipped through the *Sunday Times* sections to get the media pages. It was bliss to be in bed on Sunday morning, the duvet a sea of newspapers, the Archers on the radio and Bart in the kitchen frying sausages and tomatoes. Throughout the week she made do with unbuttered Ryvita and coffee, but on Sundays she was a sybarite and Bart was her slave.

'Shove up,' he said, coming into the bedroom with a loaded tray. She scrabbled the papers together and plumped up her pillows.

'God,' he said, sliding into the bed still holding the tray aloft, 'you're drooling. You're literally drooling.'

'You make it so delicious,' Jenny said.

While they ate he skimmed through his paper, the *Sunday Recorder*.

'It's in,' he said at last, 'page 23.' He had done a piece on a Tory councillor who had feathered his nest from as yet untraced sources. 'That'll do for the sod!'

'Don't you ever have doubts?' Jenny asked, her mouth full.

'About what?'

'Well, about whether or not what you're alleging is true? Or, even if it is, what it might do to them to be exposed?'

Bart was shaking his head. 'Not my problem, Jen. I'm a journalist, not a judge and jury. I report facts – facts I've checked inside and out. I don't pass opinions, I don't condemn, I report. So why should I have doubts?' He bit into a sausage, held between finger and thumb. 'Do you have doubts about your work?'

'No, but that's different.'

'Why?' He was hyped up now, on the offensive, and Jenny wished she had never raised the matter.

'Well, it just is. I'm not wrecking people's lives . . . no, that's not what I mean. I'm not . . . treading in murky waters.'

'Did I make the water murky?'

'No, but . . .'

'Did I make the water murky – no buts.' His voice had taken on an edge.

'Stop it, Bart, I don't like it when you start attacking me.' She wanted out of the bed and away from his unrelenting voice. Above all she wanted rid of the sausage and tomato sticking halfway down her throat.

'I never *start* attacking, Jen.' Bart's voice was silky now. 'I never ever start things . . . but if someone levels an accusation at me which I consider unfair, I make them regret it.'

She felt like a child, suddenly, standing ashamed in front of the grown-ups. As if he had read her thoughts Bart lifted the tray in one hand and lowered it to the floor. 'Come here, idiot. I can't bear it when you look like a little, whipped puppy.'

Jenny wanted to resist but it was easier to hide her face against his shoulder and slide into his arms and down the bed.

May listened to the morning service while she washed her tea cup and saucer and put them away. 'Just as I am, without one plea, but that Thy blood was shed for me, and that Thou bidd'st me come to Thee, O Lamb of God I come.'

Edward had loved that hymn, had requested it be played at his funeral. Strange that it should be played today.

She switched off the radio when the service ended and made her way upstairs. Earlier she had changed into her best nightgown, cutting the old one into pieces and putting the pieces in the bin. She fingered her knitted bed-jacket: it was clean but was it suitable? In the end she decided to permit herself one luxury, warmth. It might take a while and it would be nice to be comfortable. She debated whether or not to remove her slippers and decided they too could be allowed. She took a last look round: not a thing out of place, only bottle and glass on the bedside table, and they were essential.

Sitting down on the bed, she reached for the pills. Ten would be more than enough. The doctor had warned her about taking two together, even that could be disastrous he'd said. So ten would be safe enough.

They looked so innocent in her palm, like sweeties. Like dolly mixtures, the yellow powdery ones. She had bought dolly mixtures for Alan when he was a child. He had played shops with them, selling them to Edward in twists of paper for a bright new penny.

A shaft of pleasure went through her, as though she had been on a long journey and now was within sight of home. She put the pills into her mouth two at a time and washed them down, waiting a few moments before repeating the action. When she was done she carried the glass to the bathroom, rinsed and dried it, and put it in its place. The sun was shining in through the window and she could see

199

the results of her labours as it landed on gleaming tiles and spotless paintwork.

Back in the bedroom she lay down on the bed, composing her limbs carefully and pulling her nightdress down as far as she could. But her bare ankles worried her and she got to her feet and turned back the coverlet so that she could pull it up to cover her legs.

As she lay back on the pillow a sudden wave of nausea hit her. She had expected something like this and she tried to stay calm and breathe evenly.

It was passing when she remembered the photograph. The one thing she had forgotten, Edward with Alan in the garden, the father squinting up into the sun, the baby gazing bewildered at its father's upturned face. It was there on the dressing-table, when she had meant it to be on the bedside table so that she could look at it for as long as possible.

She managed to sit up, surprised that she could still swing her legs to the floor. Perhaps it wasn't going to work and she would have to find another way? She had wanted it to be neat for Alan but perhaps that would not be possible.

She crossed to the dressing-table, pausing half-way for a deep breath. She could see into the garden from here, the apple tree just coming into blossom, the birds flying back and forth servicing their young. She felt a momentary pang that she would not be there when apples hung on the bough, and then she remembered her objective and moved purposefully forward.

When she got back to the bed she positioned the photograph, altering it a fraction of an inch to get it right when she lay down, grateful that at last all was in order. Nausea came and went, but she kept the memory of the blossom in her mind and gradually she saw Edward there, under the tree, smiling, and Alan running across the grass on shaky two-year-old legs. 'Come on, mammy. Come on!'

Kath watched Brian's shaky progress from chair to chair. He had taken a long time to walk but he could do it now . . . just in time. In three months she would have a new baby to tote around, so if Brian could stand on his own two feet it would help.

She heard Dave at the back door, knocking his wellingtons against the step to get rid of the muck. Since he had got the allotment he spent most Sunday mornings there, trying to turn a neglected morass into a garden. It would be worth it in the end, though, if even half of his promised fresh veg came true.

'Any char?' He was in his stockinged feet now, scooping his son into the air to squeal with delight. 'Been a good lad, then?' he said, sitting in the chair with Brian on his knee.

'He has been good,' Kath said. 'I got two pages of the paper read, that's an advance on last week. Here's your tea.'

Dave was looking down at his son. 'Tell you what, lad. When I've supped this tea I'll take you for a walk. But only on one condition.' The little boy's eyes were round, as though he understood every word. 'If your mam lets me out for a pint afore dinner, that's the condition. What d'you think she'll say?'

'Tell him, Brian,' Kath ordered. 'Say, "She'll say blank off."'

'Did she say that?' Dave asked, eyes still on the baby's face.

'She did,' Kath confirmed. 'She said it very loudly.'

'Loudly?' Dave mused. 'And I was so nice to her last night as well.'

'Yes, you were,' Kath agreed. 'But she was quite nice to you, too, considering her condition.'

'True,' Dave nodded, without looking up. 'Very true.'

Suddenly Brian looked from face to face and grinned. 'True,' he said uncomprehendingly.

'God, he's clever,' Dave breathed fervently, and Kath rose to put her arms around them both.

'Off you go, you s-o-d. Keep him out while I get the veg done and the puddings in, and then you can have one pint. One pint. o-n-e.'

Dave stood up and handed her the baby. 'Take him while I find me shoes. If I can have a pint I'll do all the pots, and then I'll mind him while you read the papers cover to cover.'

201

'Promise?'

'God's honour.' Suddenly he was serious. 'I know you like to keep up with things. You've got a good brain, Kath, and I'm proud of you. You gave up a lot for me, and I'm grateful. One day you'll use that brain . . .' Suddenly he grinned, '. . . and I'll sling me pit boots and sit on me arse while you keep us all.'

'Get out,' she said and stood at the door to watch father and baby make their way down the path and turn at the gate to wave her goodbye.

'You look happy,' Toni said. Keir shrugged.

'I'm OK.'

'You've been like a cat on hot bricks for the last week, but now you look like a Cheshire cat.'

Around them the restaurant was quiet, almost deserted. Keir hated Sunday lunches and nondescript eating-places. It was always pasta, and if you found a traditional roast it was covered in gelatinous gravy, the veg still frozen in the middle. 'I'm neither particularly happy nor particularly unhappy, Toni. I'm trying to relax, actually, if no one has any objection.'

'Sorry.' She held up both hands in apology and he caught sight of the one short nail. It repulsed him, that habit of hers of keeping two-inch nails on nine fingers and biting one almost to the quick. She nibbled it when she was reading, which was almost all the time, holding her other fingers apart so that the nails stayed intact while she went chipmunk-like for the wounded one. Now she compounded the offence by reaching out that hand to cover his where it lay on the table. 'It's just that I never know what you're thinking and it gets to me.'

'You don't own me, Toni. I am still free to think.' He pushed his plate away, still half-full. 'I've had enough.'

'Me too. Shall we have another half-carafe?'

When the red wine came Keir tried to think of some innocuous topic of conversation but it wasn't easy. His mind was full of Annie Love and her invitation to him to ring. He might just bring it off! After that . . . His thoughts soared until he felt Toni plucking at his sleeve.

'You're at it again – mooning.'

'I'm just day-dreaming, Toni, nothing particularly earth-shattering. I was wondering how long that garlic has been hanging up there, if you really want to know, and whether or not it keeps its flavour.'

'It doesn't. I'm not prying, darling, it's just that I tell you everything.'

There it was again, the line he had heard in one form or another from every woman he had ever known: *I love you, why don't you love me?*' They all said it sooner or later, even someone as independent as Toni had once been. Perhaps it was him? Perhaps he sapped them of the very thing that had first attracted him? If so, the outlook was bleak.

There had only been one woman who hadn't been like that, or one girl – Jenny. She had been so sweet the night that Jed died. And then she had taken fright and given him the brush-off. If nothing came of the Annie Love thing he must look Jenny up. She was in London, Alan had told him so. In the mean time there was the problem of Toni – if he didn't repair fences she'd be weeping by nightfall, and he couldn't cope with female weakness, not with so much to get clear in his mind before he made that phone call.

'Tell you what,' he said, 'why don't we skip coffee and go home for a siesta?' Her eyes positively lit up, poor bitch.

'I've heard about northern men on Sunday afternoons,' she said.

'It's all true.' He reached to pull her out of her chair. 'Let's go home and I'll prove it to you.'

Elaine was on the phone in the hall when Jenny came through the door.

'Please, Nico. Please.'

Jenny put a finger to her lips and then pointed upstairs to indicate she would go on up, and Elaine nodded, but as Jenny mounted she couldn't fail to hear Elaine's half of the conversation.

'Oh please, Nico. Half an hour, just half an hour? I know, I know . . . I'm not going on about it, it's just that I need to see you.'

In Elaine's flat L.J. sat placidly on the floor, surrounded by an assortment of dolls and soft toys. She looked up at Jenny and smiled.

'Hallo,' said Jenny, taking off her coat. 'Are you having a tea party?'

The little girl nodded solemnly and held out a fluffy rabbit.

'Thank you.' L.J. was enchantingly pretty, Jenny thought, settling on the floor beside her. She had her mother's fragility and her father's features – disconcertingly perfect nose and mouth, neat but determined chin. The eyes were Elaine's, though, huge and somehow sad, even in the baby face. What was Elaine doing down there in the hall, grovelling to some man?

L.J. had picked up a doll and was pressing her to the rabbit. 'Love you,' she said, and then again, 'Love you.' Jenny smiled and nodded encouragement.

Elaine came back into the room, looking sheepish. 'Hi, Jen.'

'Who was that?' Jenny said, unable to pretend indifference.

'A guy. Nico. He comes in the pub sometimes.'

'Is he nice?'

Elaine shrugged. 'Is any man nice? He's OK.'

'Well, if he's only "OK", Elaine, why were you practically begging down there on the phone? You . . . you're so . . . well, you've got such a lot going for you. You're pretty . . . I'd say you were beautiful. You've got talent and personality . . . and you were *grovelling*!' She wondered if she had gone too far but Elaine was smiling.

'Oh Jenny, it's so simple for you, isn't it? Never lose your dignity, that's the message. But I haven't got any dignity, I lost it a long time ago, if I ever had any. Probably I didn't. I need a man, Jen. I'd like one on your terms but I know I won't get it, so I settle for what I *can* get: someone now and worry about tomorrow tomorrow.'

Jenny was opening her mouth to reply but Elaine held up her hand. 'Not now, Jen. It's lovely to see you, so let's not spoil it. I've got some vodka and two hours before I go to work. Let's just talk. If it'll please you I'll go to Jersey one day and meet a millionaire like Barbara, and then we'll all be happy. Now, where's the glasses and tell me why you're not out with Bart on a lovely sunny Sunday evening.' She moved to the dresser, her thin body appearing to bend like willow as she walked, pausing on the way to stroke her baby's hair from her eyes.

'Bart is off on one of his mysterious assignments,' Jenny said, scrambling up on to a chair. 'And I wanted to see you and L.J. She really is turning into a beauty, isn't she? We've got to get you together with Kath soon . . . maybe for the christening.'

'That's what Alan said when he came round,' Elaine said, smiling. 'He doesn't change, does he? I mean, I hadn't seen him for ages but we just picked up where we left off.'

'Where we left off,' L.J. mimicked and then laughed aloud at their startled faces.

Steven switched off the set when the news ended, conflicting emotions within him. The Royal Marines had recaptured South Georgia and as he had watched John Nott, the Defence Secretary, and Maggie Thatcher standing outside No. 10 he had felt a savage satisfaction that Britain had shown its mettle. Even when she had put in her two penn'orth he had found himself nodding. 'Let us congratulate our armed forces and the Marines. Rejoice, rejoice!' Maggie had said, and he had almost cried out, 'Hallelujah!' until he remembered who she was and scowled his contempt.

'They've done it,' he told Cissie when she came in from chapel. 'They've taken South Georgia.'

'Very nice,' she said, which was hardly adequate, he felt. She got on with the fry-up straight away, tipping the veg from dinner-time into the frying pan and stirring them into a hot, spitting mass. He loved bubble and squeak even better than freshly cooked vegetables. The puddings were better hot from the oven, but when Cissie reheated them on a baking tin over the fire they tasted all right, well salted and dipped in gravy.

'It's a bit since our Keir was home,' she said as they ate

from plates on their knees, while a female singer pranced about on the telly in a frock like a rubber glove.

'Aye, well . . . he's got his way to make.'

'D'you think he'll ever come back?'

'Mebbe. Mebbe not.

'Steven,' she said, pushing aside her plate, 'do you still love me?'

'What a bloody silly question for a Sunday night. No, as a matter of fact I've gone off you.'

He looked at her expecting a grin but Cissie looked a bit like a lost bairn. He held out his plate.

'Fill that up and I'll think about it. I don't know, a fifty-five-year-old woman asking a question like that.' But while she refilled his plate he realized that it was only after fifty that you began to have doubts. They had that in common. When she held out the loaded plate he gripped her arm above the wrist. 'Of course I love you, you silly old woman. I'll prove it to you later on if I get a bit peace now. Turn the telly up and stop wittering on. And while we're on the subject of our Keir, I've been thinking about the phone . . .'

It was half-past ten when Jenny accepted that Bart would not be home that night. She put aside the book she was reading, *Tess of the D'Urbervilles*, bought after she had seen Roman Polanski's evocative film version, and went to run a bath. She was just about to step in when the phone rang in her bedroom.

'Hallo, is that you, Jenny?' She had expected the caller to be Bart, and for a moment the woman's voice failed to register.

'Barbara?'

'Yes, it's me. Sorry this is so late but I'm off back to Jersey in the morning and I couldn't go without ringing you. I hoped we'd get together while I was here but it'll have to be next time.'

'How are you?'

'Blooming. And Guy is wonderful. Unbelievable! I've got a *huge* ring. Guy says it has to last a lifetime so it's an investment really. What about you?'

'Oh, I'm fine. Working hard.'

'Still with your newspaper magnate?'

'If you mean Bart, yes.'

'I always thought you'd wind up with one of the Durham boys. Is Euan still in Africa? I know Keir's doing well, mum says she's seen his name on TV several times.'

'Yes, he's still with Eastern. I think he's a PA . . . a production assistant now. And Euan is still in Zimbabwe but not very settled. He says he'll be back this year.'

'Do you still write to him? You'd do well for yourself there, his family are rolling!'

'I'm with Bart, remember?'

'Oh . . . at least you have something in common. I saw Alan's mother, by the way, and told her how much I admired her. I mean, all the sacrifices. I said he still had a long way to go, of course she must know that, but I think they've both been marvellous. Guy says it takes years and years, but she seems quite resigned to it. Amazing really. She's not very friendly, though, is she?'

'I've never met her. But in fact Alan's doing quite well. I've seen him in court and he's had some articles published in national papers, so he'll have been quite well paid.'

'Oh, well . . . she's in for a pleasant surprise, then. Anyway, I just wanted to tell you that all's well and you'll be very welcome if you come to Jersey. Aunt Bea still talks about you.'

'Are they both well?'

'Fine. Coining it. Oh, I had tea with Kath, by the way, in Dunn's. Poor Kath. Still she made her own choice. It wouldn't do for me.'

When Barbara finally rang off Jenny went back to the cooling bath and had a quick scrub instead of the long soak she had promised herself. Had she ever been close to Barbara or was that a figment of memory? Anyway, there was little or nothing there now. It had been like talking to a stranger.

Jenny thought of what she had said about Bart. Did they really have something going? It had been nice to be able to say they had, anyway, and sweet to be able to put her right about Alan. Whatever Barbara had said to his mother, it didn't matter – she would find out soon enough that her son was destined for great things.

19

3 May 1982

Jenny was glad of Kath next to her in the pew, both of them dressed in the nearest they had to mourning.

'It brings back memories, doesn't it?' Kath whispered as they hunted for the first hymn, and Jenny nodded. 'Just as I am without one plea . . .' She joined in once she was sure of the tune, singing as loudly as she could. There were only a handful of people in the church and she didn't want Alan to be hurt by the sparse attendance.

Her father had phoned her with the news. 'Did your Alan live in Stanfield Gardens?' Jenny had smiled at his use of the possessive pronoun and said yes. 'Then I think his mother's died. It's in the paper. "Body found in suspicious circumstances" – well, it doesn't say "suspicious" but it doesn't say anything else either, so something's up.'

She had phoned Alan's London number straight away but the man who answered had confirmed the bad news and told her that Alan had gone north. She had tried to keep in touch over the next few days and then she had come north herself to give what support she could.

'I went over as soon as I heard,' Kath had told her that first night. 'He's so composed it makes you want to weep. He's got no one now – think about it, no brothers, sisters, cousins, uncles, aunts. I curse our lot but at least they're family.'

'He's got us,' Jenny said.

'Yeah, but that's only two and a half if you count me bump. Who else is there who could come and support him?' They had ticked them off one by one and decided that Keir was the only hope. 'I know he's a shit but he's a man,' Kath had said. 'Men like other men round them when they're up against it. Will you ring him or shall I?'

But Keir had had urgent business. 'Honestly, Jen, at any other time I'd've come like a shot but right now it's out of the question. I may be leaving Eastern to work on a mega new programme for Global, and I have to nurse it, I really do. I'll write, and I'll be in touch as soon as he's back here, but I can't come up. Sorry.'

Jenny had put down the phone disappointed but not surprised, and had petitioned her parents.

'We'll come,' her mother had said. 'Dad can take an hour off. Alan's a friend of yours and, besides, I don't like to hear of such a story, friend or not.'

The inquest verdict had been 'Suicide while of unsound mind', and the clip in the local paper had been brief. 'Not sensational enough,' Kath had said, 'thank God.' And Jenny had agreed.

Now they stood behind Alan, alone in the front pew except for an aged and distant relative, and sang the hymn May Denton had requested in her final letter to her son: 'O Lamb of God, I come'. And then the vicar was rising to make his address.

'It is not given to everyone to be great. Most of us live out our small lives in an unexceptionable way, doing our best, content to be at peace with our neighbours, do right by our families, render what small service we can to God. May Denton was an ordinary woman. She did not seek the limelight. Some might say she shunned it. But she knew where her duty lay, in service first to her husband and then to the son she raised to be a credit to her and to his profession.'

Jenny looked sideways and saw Kath's jaw tighten. She wouldn't like Mrs Denton's being termed a handmaiden and that seemed to be the vicar's intent.

'So I say to you, of such small people is the Kingdom of Heaven, and let there be no doubt that May will have her place there. Whatever drove her to that last, sad act . . .'

'Shut up, shut up,' Jenny said under her breath, seeing Alan's shoulders tense under the dark suit. He had broadened out in the last two years, the neck beneath the dark head was muscular and strong, and he seemed to tower over her more now than he had in the old days. If only Keir had been there with him, or better still, Euan. Jenny dropped to her knees at the vicar's invitation and prayed silently that Alan might be comforted.

'Will there be enough?' Jenny asked anxiously when they were back at the house and in the cold kitchen, and Kath grinned.

'One plate would feed that lot. There's only seven counting us, Jen.'

They had taken the funeral tea out of Alan's hands, swearing they knew all about such things and he did not. He had insisted on giving them twenty pounds but they had spent less than six and the change lay on the spotless window-sill.

'It's a bit bleak in here, isn't it?' Kath said as they waited for the kettle to boil. The walls were painted duck-egg blue, the paintwork brown; the working surfaces were well-scrubbed wood, and the sink was earthenware and low down.

'God, it's made for a pigmy,' Kath said as she bent to rinse a milk bottle.

'I think it's as it was when the house was built,' Jenny said. 'It looks like a 1920s or '30s house, and this is what the kitchens were like then.'

'I thought me mam's was a bit primitive but this is something else,' Kath commented.

'It's very clean,' Jenny answered defensively.

'It gives me the creeps,' Kath said. 'Poor Alan. Thank God he's got London to go back to.'

'He's got company where he lives,' Jenny said. 'I think there are two other barristers. I spoke to one on the phone, and he sounded nice.'

'I'm glad your mam and dad came to the church,' Kath said as they put milk into seven cups. 'I didn't ask mine. If she'd thought I wanted her there she'd have said no on principle.'

'No thaw?'

'No. Not even with the bairn. She's all right with him but there's no warmth like she has for our Kevin's bairn. She sees Brian as the little hiccup that spoiled her plans, and she can't forgive him.'

'She'll come round in time.'

'Time's running out, Jen. Dave is getting fed up. He tried hard at first, but now he's getting a bit bolshie. I know he's in the right, but it still hurts me. It's a terrible thing, what you feel for your mother. I hope I never do anything like that to any child of mine.'

'I'll take the tray,' Jenny said, eyeing Kath's swollen waistline. 'You bring the sandwiches.'

The living-room had the same ancient feel to it. Black-framed photographs of the Lake District hung on the walls, with an embroidered picture of a crinoline lady in a cottage garden. But the heavy dark furniture gleamed with polish and the curtains hung crisp at the window. May Denton might not have updated her house but she had certainly cared for it.

Three neighbours were bunched together on the settee and the elderly cousin sat in an upright chair with Alan at her side. He flashed at Jenny a glance of gratitude tinged with desperation, and helped to hand out the tea.

They had all been offered sherry when they arrived. Now Jenny collected the delicate glasses and carried them out to the kitchen, trying to imagine Alan as a child in this weird time-capsule and marvelling that he had grown up to be such a normal teenager. He had always been more reticent than the others, but she had never imagined his background to be so austere. There was no hint of frivolity here, no sign of wear or tear. Everything was exactly in place, without any evidence of the normal trivia of daily life.

She turned from the draining-board at the sound of the kitchen door. It was the elderly cousin, bearing her cup for a refill.

'I was going to go round with the pot,' Jenny said, but the old lady held out her cup to be filled there and then.

'It's very sad,' she said. 'He's all alone now.' Jenny added milk and turned to refill the kettle. 'Poor May, she got nothing out of it in the end.' Jenny tried to look politely interested but her mind was flicking to the other room and Kath coping there alone.

'Everyone warned her, but she didn't listen. We said it would end in tears, and it did. I suppose you could say she's had the boy, but he's been a burden to her by all accounts, with a fancy education and that kind of thing.'

'He did it on grants and scholarships,' Jenny said, feeling piqued. 'He's terribly clever.'

'So was his father – too clever for May. We said it was funny at the time . . . May and December.'

'Jenny!' Kath's face in the doorway was imploring. 'Where's the hot water?'

Jenny filled the hot-water jug and they all went through to the living-room.

The neighbours were talking now, unleashed from inhibition by tea and Dundee cake.

'I did try,' one said, shaking her gold ear-rings, 'I even called the very day. "Can I do anything?" I said. I mean, I'd have done anything for her. She was a nice neighbour.'

'I'm very grateful to you,' Alan said gravely. 'And I'm grateful to you all for coming today. It helps.'

'I wonder if he's cried?' Jenny thought. Could Alan cry? *'If I am wounded do I not bleed?'* That was Shakespeare, but it seemed apt. She hovered in the background while they took their leave one by one and then joined Kath to do the washing up.

'I'm making pannacalty for your dinner,' Cissie said encouragingly.

'Aye. Nice.' Steven kept his eye on the television, hoping for more news of the *Belgrano* sinking. The Argentine cruiser, the *General Belgrano*, had been sunk the day before and hundreds were dead. More important, the knives were out for Maggie. Had the ship been inside or outside the exclusion zone? He felt uncomfortable about it, hence his thirst for up-to-date news. On one hand there was only one way to fight a war: clobber the enemy. Never mind which way the bugger was pointing, flatten him. On the other hand it irked him to have anything in common with Maggie Thatcher, so if it was proved she had sunk the *Belgrano* for her own ends, he could salve his conscience.

Outside the sun was shining but he'd lost interest in walking since the dog died. There didn't seem much point in wandering round the town like a loony, you needed somewhere to be going and he was sick of the beach and sicker still of the Aged Miners' cottages. They all addressed him as an equal now, men who'd been long in the tooth when he was a shivering trapper-boy. He still had a job – only just, but it was still there – and they should remember that.

Behind him Cissie shut the oven door with a clang. Something was coming! She had clashed the bread board down to slice the taties and clouted the bacon and black pudding into the pan. He shifted uneasily in his chair. Now what?

'I've been thinking.' So it was as bad as that. Cissie's thinking usually meant more HP or a visit to her brother in Scarborough.

'Aye.'

'For God's sake say something else but aye. You're getting glakey, sitting in front of that thing all day like a Buddha unless you're down the pit. What you want's another dog.'

So that was it. Steven put out a foot and tried to touch his boot but it was irritatingly out of reach and in the end he had to get up and bend for it. 'There's no dog coming in here.'

'Why not? They've got some lovely Jack Russells down the street . . . little pictures.'

'Talk sense, woman. Another dog at our age! And if I was having a dog it'd be a dog, not a mouse on stilts.' He finished tying his boots and made for the door, but her tongue was ahead of him.

'That's it, run off! Typical man – can't stay and discuss, oh no. It's my house as much as yours and I can have a dog if I want.'

'Have one. Have two and breed them, just don't drag me into it. I never wanted the last one. That was your doing.'

He knew what Cissie was up to, Steven thought as he went down the yard. She wanted a dog to get him out of the house. But he was past training a puppy. He'd like a dog well enough – a head on his knee, a wet nose at his hand – he'd like the dog back like he'd like his son back, to give a bit meaning to life. But there were stages, and that

213

stage was past. He turned the corner and saw the sea, blue and vast. Too vast. He turned on his heel and made for the Half Moon, fishing in his pocket for the price of a pint.

Keir had negotiated the menu easily enough, but the wine list was different. Should he be indifferent, as though above caring, or plead attractive ignorance? In the end he chose the latter course. 'I know nothing about wine, I'm afraid. Shall I use a pin, or will you choose?'

Annie Love picked out a wine, reeling off its complex French name, and then turned her attention back to Keir. 'Now . . . tell me more about this programme idea of yours.'

Keir tried to remember what Toni had told him about the proposed pilot, loading his words with as much jargon as seemed wise. He must seem *au fait* with big-time television, but not cocky. Annie Love had a reputation for eating people too big for their boots.

'Are you married?' she said abruptly, cutting across his flow.

'No. Not even involved.'

She had taken an olive from a dish and was shredding it slowly, keeping it between her front teeth rather as a half-back dribbles a ball. She appeared to be lost in thought and Keir wondered if he should stay quiet or go on.

'You come from the north-east?'

'Durham. Is it in the voice?'

'A bit. Not much. How long were you at Oxford? What did you do there? What made you choose television? What did you mean to do when you went up to Oxford?' The questions came pitilessly while she demolished two more olives and sipped her gin.

'I was interested in politics, then. I still am, but more as a spectator now. My father's an active party member, Labour obviously . . .'

'Why obviously?' She was on to him like a flash.

'He's a miner.' Keir said it simply with just the right amount of pride in the voice. Was she impressed or put off by his origins? God, the bitch was deeper than the mid-Atlantic. She had on a petrol-blue sleeveless dress and her arms were thin and muscular, her breasts no bigger than two half peaches, with nipples like small plums. Keir found he was imagining what it would be like to take that well-muscled machine to bed, and the thought disconcerted him.

'Good,' she said suddenly. Just 'good', nothing more. And then the food was arriving, pink salmon mousse with a spray of greenery to set it off, and the wine was being poured.

She raised her glass. 'To a Durham lad,' she said, drawing back her lip over her teeth as she smiled. Her hair was the colour of autumn leaves and her long, unpolished nails curved in slightly at the tip. If he could take her to bed they could sod the programme, and that was a fact.

'Will you sell the house?' Jenny asked when she and Alan were sitting quietly together in the front room. Kath had gone to catch her bus to Belgate, and Jenny had made more tea and carried it through to the silent room.

'I don't know.' Alan looked around. 'It's my home, I can't imagine anyone else living here.'

'It's a nice house,' Jenny said. 'The garden's lovely.'

'My father made the garden.' Alan leaned to pick up a framed photograph. 'That was him, he lived here before my mother married him. I did a bit of gardening when I was at home, and mother pottered on.'

'You always call her mother.' Jenny couldn't resist saying it.

'I know.' Alan looked self-conscious. 'Dad called her that, I think. I only dimly remember him but he used to say "your mother" – "Ask your mother" or "your mother will fix that".'

'It sounds as though they were happy.'

'I think they were,' he said slowly. 'I'm not sure. My mother', he grinned self-consciously at the word, 'would never talk about it . . . about their life together. But she cherished my father's memory. She had great respect for him, I know that. It was never said but I got the feeling that I'd never quite measure up to him.'

Jenny nodded, trying to think of something else to talk

215

about. She couldn't leave Alan here in this cold house unless she tried to divert him a little.

'More tea?' she asked.

He held out his cup. 'I heard from Barbara. Typical Ba-ba-black-sheep: three lines of condolence and twelve of how she's going up in the world. And Keir wrote, too.'

'What did he say?' Damn Keir, she thought. Why did she still want to know what he said?

'He's doing well, and says he'll contact me in London. He's still with Toni but is changing jobs. I get the feeling that Toni is about to become redundant, you know Keir.'

Jenny felt her cheeks colour. Was that an innocuous remark or did he mean just what he said? 'You won't have heard from Euan?'

'No. Does he know?'

'Kath wrote to him, but it takes a week for a letter to get there.'

'I got a card from Elaine, with a cross on it from L.J.'

'She's wonderful. I'm not the maternal type but I could eat L.J. She's Elaine with a little flesh on the bones.'

'She takes after her father, surely . . . that lively little face?'

'Oh yes,' Jenny said carefully, 'yes, she's like her father.'

'We can look out for fireworks, then, when she grows up.'

Again Jenny nodded and let the subject lapse.

'I'm seeing Kath tonight,' she said when it was time to leave. 'Just in the Belgate pub for a drink. Do you want to come? We're only going to talk about old times, and you'd be welcome.'

'Thanks, but I think I'll stay in tonight. I've got a few things to see to and a great deal of thinking to do. I'll be in touch tomorrow.'

He was bending to kiss her cheek, his lips like the touch of a feather. 'Thank you, Jenny. There's no one I'd rather have had here with me today.'

Keir lay with his eyes closed, the picture of Annie's face above him still there in memory – her mouth open but the lips somehow pursed, the eyes open and raised. God, it

216

had been good! She had done it to him, no other word for it. He had never had that happen to him before, and he liked it.

He could hear her in the shower and he drew himself up on the pillows to be ready if she wanted more. But when she came into the room she was already dressed in a black cocktail gown with a low back, slipping her bare brown feet into strappy high-heeled sandals.

'I've got to dash. Feel free to use anything. Slam the door when you go out. I'll be in touch.'

She was walking out on him! Keir tried to think of something to stay her but she was slipping ear-rings into her lobes, transferring things from her huge day-time bag to a shiny black pouch.

'See you,' he said as she made for the door, and cursed his inadequacy long after she was gone and the door closed behind her.

He had been OK an hour ago, though. More than OK. He flexed his various muscles cautiously to see if there were any after-effects. None. Good! The sheets were some kind of silky material, gunmetal grey, and the bed was huge. He wriggled down, trying to reach the end with his toes. He was six feet, so the bed must be seven. He slid round, measuring it across – as wide as it was long! – and tried to calculate her salary. Big . . . but this big?

He lay for a while and then threw back the covers. The carpet was silky under his bare feet and when he touched the sea-blue walls they were textured in some way. He didn't actually like the décor but he was sure of one thing: nowhere else in Britain was there a room anything like this. Exclusive. He liked it.

In the bathroom grey and blue towels were stacked in gunmetal shelving, and one wall was mirrored from top to bottom. He looked at himself, hands on hips, and then turned, squinting to see if her nails had marked his back. Toni would go spare if they had.

His back was reddened but otherwise OK, and he bent to turn on the bath taps. Sod Toni, anyway, she didn't own him. She would have to be reckoned with, though. He couldn't just walk out, not after two years.

Lying in the scented water, Keir tried to work out what

came next. What if he never heard again from Annie Love? Perhaps this afternoon had just been a quick fling, something she did with all aspiring PAs? After all, why should he be different? But in his heart he knew that he was, that his father had been right all along to tell him he was destined for something big.

He noticed a cord dangling near his head and reached to pull it. Music filled the bathroom. Vivaldi. Jesus Christ, this was the life. He heaved himself from the water and went in search of the bottle they had broached an hour ago, carried a brimming glass back with him, and held it aloft while he climbed back into the water. 'Not bad for a boy from Durham,' he thought as he drank. And more, much more, to come.

20

14 June, 1982

It sat there, unexpectedly green but recognizably a telephone.

'Well go on, try it,' Cissie said when she had shown out the telephone engineer.

'I'm not touching it,' Steven said. 'It's your phone.'

'Ring our Keir,' she coaxed.

'At this time of day?' Steven shook his head sadly. 'He's working, woman, use your brains. Better still, get down the shops and get the messages in, and by the time you get back somebody might've got wind and they'll ring *you* up. That way it costs nothing.'

He waited till the yard door closed behind her and then lifted the receiver. It was burring, so it was all connected up. He tried to think of who to ring. No one who would tell Cissie, and that meant no one in Belgate. In the end he settled for the speaking clock, getting the number from the card the engineer had left and listening to the metallic tones add on the seconds for quite a few moments before he said 'thank you' and put down the receiver.

He switched on the TV to check up on the war. Everything was going nicely if you believed all they told you. He tried to concentrate, even switching his thoughts ahead to the evening when they might catch their Keir in and get a bit conversation. It was no use. In the end he rang up the Labour and Conservative clubs, asking each time for a fictitious character, and then rang directory enquiries to ask for his own number. They knew nothing about it, which filled Steven with rage until he remembered you had to allow for bureaucracy. They would have it soon enough, and, any road, Keir would have it by tonight and that was all that really mattered.

Kath threw the paper aside and struggled forward in her chair to mend the fire. The room was too warm but she needed the hot water the back boiler provided. Brian sat inside his lobster pot playing placidly enough until he saw her move.

'Up, up, up,' he chanted, hauling himself up on the edge of the playpen. He wanted to be picked up and he was sure to need changing, but Kath couldn't be bothered, not yet. She was tired. There was nothing done yet for Dave's meal, a mountain of washing awaited her, and she had an itchy rash in what seemed like every conceivable fold of her skin.

'Not long now,' the midwife had said this morning. But 'not long' was another month and Kath wasn't sure if she could stick it.

'Up, up, up,' her son chanted, almost the only word he knew other than 'dada' and 'cars'. She ignored him until a brick flew through the air and clipped her eyebrow.

'Stop that, Brian. That's naughty.' A tear formed in the corner of her eye and trickled slowly down her cheek. Was this what her mother had meant? Was she sinking slowly into a morass of self-pity? She wanted her mother now, wanted her badly, but want would have to be her master.

She heaved herself out of the chair and bent to pat her son's head. 'That's a good boy, pet. We'll have a game in a minute when mammy's had a rest.' She switched on the TV and lowered herself back into the chair. A magazine programme was in progress, with an elegantly dressed royal-watcher seated on a settee talking to an interviewer. As usual Kath wondered if it was one of Keir Lockyer's programmes. It gave her a thrill to see the name of someone she knew roll up in the credits. In the beginning she had even bought the *Radio Times* to see Jenny's name tucked away. 'They've done well,' she thought. 'They've made something of themselves.'

Uncomfortable, she turned her attention to the TV.

'The Princess of Wales' room awaits her in St Mary's

Hospital. There are sure to be flowers there already, in case of a hurried arrival. The room has probably been decorated in colours the Princess is known to like, and staff will have been hand-picked, everything to make sure the Princess's first delivery is as easy and comfortable as possible.'

'Tell us about the royal nursery,' the interviewer said.

'Not now,' Kath said, levering herself out of the chair. 'Not sodding now, thank you.' She switched off the set, sobs shaking her chest, but whether of pain or anger she didn't know. In the playpen the baby started to cry, and, contrite, she picked him up and cuddled him.

'Never mind, pet, don't cry. It's only mammy being silly.'

Keir put a few essentials into a travelling bag and carried it through to the living-room, hiding it behind a chair. He looked at his watch. Toni would be here any moment: all the same . . .

He dialled Annie's number, reaching to turn down the volume on the TV as he did so.

'Annie . . . it's me. I should be with you in an hour. No, not yet, she isn't back. Yes, I'll cope. No . . .'

At the other end of the phone Annie Love was interrupting him, something about the TV news. He leaned forward and turned up the volume. 'A white flag is flying over Port Stanley . . .' So it was over!

He turned back to the phone. 'Yes, I got it. See you later. Love you, Love.'

He was still listening to the announcement when Toni let herself in at the door.

'Hi. Want a drink?' He poured her a gin and tonic and got straight down to business. 'Toni, there's no easy way to say this. I'm leaving. It's not that I don't care for you; I do. But I think we both need a breathing-space . . .'

Her head had dropped at first, but now she looked up, twin spots of colour on her cheeks looking odd among the freckles. 'It's Annie Love, isn't it? You're screwing her, you bastard. When you got the job I wondered. She heard of you from someone, you said. I bet! But not about your

221

ability in a studio, Keir, because you haven't got any. You're strictly a desert in that direction.'

'There's no need for this, Toni. Let's be dignified.'

'No, let's not be dignified. You utter bloody swine! When I think of the way you've made use of me, my flat, my contacts . . .'

The phone cut her off in mid-sentence and they both looked at it. In the end it was Keir who lifted it, handling it gingerly for fear of who it might be. Apprehension turned to amazement and then to irritation.

'I'm glad, dad, you should've done it years ago. But I can't talk now. I'll have to ring you back.'

'I don't like theatre critics,' Jenny decided, looking through the glass into the studio, where one of the species was recording his opinion of a new play. He had the oleaginous look of a fish-oil expert as he set about tearing the piece to ribbons. Jenny thought of all the effort that must have gone into the production, the writing, the casting, staging, rehearsing . . . and one man was killing it off with a few phrases.

She felt increasingly uneasy about her role in the new arts programme which all too often seemed less a celebration of the arts than a crucifixion. Nevertheless she thanked the critic for coming as she took him down to Reception and chatted with him until a car arrived to whisk him to his next killing-field.

In the canteen, where she drank coffee, everyone seemed to be talking about imminent victory in the Falklands, so that she felt a little ashamed of her own preoccupation with personal affairs. Back in her office she dialled Bart's office number.

'He's not in yet,' said the girl who answered. 'He'll be in later, though . . . he's not on assignment at the moment.' Which meant that the girl was wrong or that Bart had been lying when he had said he was. Jenny dialled his flat, not really expecting an answer, but the receiver at the other end was lifted at the third ring.

'Hi.'

'Hi,' Jenny said, taken aback at the sound of a woman's voice when she had expected Bart's.

'Can I help?'

'I wanted to speak to Bart. Is he there?'

'No.' The girl sounded more cautious, less confident now. 'I'm afraid he's not. I'm just here . . . to collect something. Can I give him a message?'

'No thank you,' Jenny said, 'I'll ring later.'

When she had put down the phone she sat and considered. He might be at his club. On the other hand, did she really want to speak to him at this moment? Probably not. She was about to start on the running order of next week's programme when the phone rang.

'Jenny?'

'Elaine! Are you OK?'

'Yeah . . . I'm sorry to ring you at work. Will you get wrong?'

'No, of course I won't. It's nice to hear from you. I was going to get in touch with you, anyway.'

'Well, it's about tonight. I'm in a jam, Jen. Bernie and Carol both have dates, and I've got a gig. Not at the pub, a proper gig. And I've tried the sitter, but she's already booked.'

'I'll come,' Jenny said. And sod Bart if he turns up, she thought. 'What time do you want me? I'll come early so you can get ready in peace, and we'll have a natter before you go.'

'So it's over, then,' Elaine said when Jenny arrived.

'Yes, thank God. If I never hear the word Falklands again it'll be too soon.'

'They were brave, though,' Elaine offered.

'Yes,' Jenny agreed, 'they were brave. I only hope it was worth it in the end.'

'Right,' Elaine said, when they were settled with tea, 'I've got five minutes before I need to get ready. Let's have the news.'

'There isn't much,' Jenny said. She couldn't tell Elaine about Bart, not till there was something definite to tell. 'Kath's fed up, which is not surprising, and I think Alan's avoiding me since his mother died, I don't know why.'

'He's probably just licking his wounds,' Elaine said. 'I

get like that when I'm unhappy, I don't want anyone or anything for a while. It wears off eventually.'

'You look pretty tonight,' Jenny said. 'Not to change the subject, but you do.' Elaine's hair was piled on top of her head and she looked healthier than Jenny had ever seen her look.

'Things are going well for a change,' Elaine said. 'If this gig goes well tonight . . . well, it's important so thanks for coming.'

L.J. was suddenly standing in front of Jenny, wispy curls pony-tailed, a striped T-shirt under her dungarees.

'Story, Jenny.'

'Story what?' Elaine said, prompting her daughter to say 'please'.

'Story *now*,' L.J. said, beaming delightedly from one to another.

'Come on, then,' Jenny said, 'climb up. Which story do you want?'

They went through *Babar the Elephant* before Elaine left. 'Don't let her take the lend of you, and there's some ham and pease pudding in the fridge. Help yourself to drinks. I'll keep the cab when I get back so you can get straight away.' She stopped in the doorway and retraced her steps. 'Ta, Jen, I don't know what I've done to deserve you. I'll never pay you back.'

'Yes, you will,' Jenny said balefully. 'I want a mink out of your first platinum disc.'

She was reading Postman Pat to a sleepy L.J. when the bell rang downstairs. She went out on to the landing, the child in her arms. A man's figure was visible through the glass. It couldn't be Bart, surely? He didn't know she was here.

She tried to put L.J. down but the child clung and in the end it was easier to carry her down the stairs. Jenny opened the door, her heart thumping a little until she saw who was standing there, the auburn hair still unruly atop the six-foot frame but the white skin tanned now, making him look like a film star.

'Euan! It's not you.'

He was looking at his arms, feeling his chest, looking puzzled.

'I think it's me,' he said solemnly. 'It was when I got to Heathrow.' And then he was hugging her, shifting parcels and champagne in his arms, looking at last at Elaine's daughter.

'So this is the famous L.J.,' he said. And then, more gently, 'Hallo, princess.' He held out his arms and to Jenny's surprise L.J. went to him while she scrabbled for the parcels. Euan held her close to him and looked over her shoulder to Jenny.

'She's so like Jed,' he said. 'So very like Jed,' and his eyes filled.

Jenny felt the weight of the lie descend upon her more heavily than ever, but there was no other option but to precede him up the stairs, agreeing with every word he said.

21

20 July 1982

There was the usual air of mild grief that always attends the making of a pilot programme, Keir thought. Annie Love was standing with her back to the studio floor, as though disowning the whole project, and Jeremy Aird, who was to present, was pacing the floor drinking coffee from a paper cup. Keir watched him enviously, thinking what he would give to get into that brightly lit arena and move easily, mike in hand, from one member of the audience to another.

'What d'you think?' the researcher said morosely. She was a small, dark, flat-chested girl with a hook nose and a man's hair cut who resisted any attempt to 'put on her'. Keir had tried alternately charming and bullying her, and as neither had worked he was trying to ignore her. Now, though, he followed her eyes to the presenter's long lean figure in the Italian suit that must have cost a month of even the enhanced salary Annie Love had allotted him.

'What d'you mean?' He tried to sound lofty.

'Him . . . Aird.'

'I know who you mean. What about him?'

'He's pissed.'

Keir winced at her tone more than her words. What a little crud she was. All the same, as he watched the tall man lurched slightly, putting out a hand to a camera to steady himself, and a tiny bead of excitement formed in the corner of Keir's mouth. He wiped it nervously away. What did he do with this?

'All right, everyone.' The floor manager was holding aloft his running order to attract attention. 'Settle down, please.' Annie Love was moving towards Jeremy Aird, and shaking her head. She turned, searching past cameras and crew until she found Keir. He obeyed her imperious wave.

'Get him out of here,' she said. 'Don't let the audience see anything's wrong. Make sure he's got no booze in his dressing-room, and then come back here . . . We've got to salvage something.'

'I'm all right, Annie.' Aird was shrugging off Keir's hand. 'You know me, I need one down the hatch to set the old cogs in motion.' But she was already walking forward into the lights, flashing her toothy smile, moving like some feline machine now that she was on show.

'Ladies and gentlemen, my name is Annie Love and I'm executive producer of this series. Your producer, Graham Cammock, is up in the gallery where all the real work is done. I was going to do what we call a "warm-up" – that means I tell some slightly risqué stories to set you all laughing merrily, so that the chairman of tonight's debate finds you all affable and charming when he takes over. That's the theory. In practice it hardly ever works out like that and tonight is no exception . . .'

Keir was standing, lost in admiration of her nerve, until he heard Jeremy Aird burp softly and then again, more loudly. He gripped his arm and turned him away from the cameras.

'Come on, chum, you need a drink.'

The dressing-room was spartan, a fact Keir still found surprising. He had expected star dressing-rooms to be sumptuous, but they were all the same: cord carpet, textured walls, a couch and a vanity unit, a shower in the corner. Only the mirror with its encircling lights was in the least glamorous. He found a half-drunk bottle of Glenlivet in Jeremy Aird's travel bag and poured a generous tot into a water glass.

'Drink this and for God's sake don't say I gave it to you. Then lie down . . . that's it . . . take your jacket off. I'll be back for you in ten minutes.'

Keir waited till the glass was drained, then rinsed it and put it back in place. He poured half the remaining Glenlivet down the sink and capped the bottle. 'I'll have to take this, Jeremy. Annie's orders.'

'Tell her I'm tired,' Aird said, already sinking on to the

227

couch. 'I just got off a plane from Zurich for Christ's sake, what does she expect?'

Keir carried the almost empty whisky bottle back to the set, folded inside a copy of the *Herald Tribune* he had found on Aird's dressing-table. He could hear gales of laughter as he got near. Annie certainly knew how to pull an audience. She finished up with an apology for the delay and promised speedy progress.

'How is he?' she said, when she was back by Keir's side.

'Drunk. Quite comprehensively from the look of it.' He moved the paper and let her glimpse the bottle.

'God damn the man. I warned him. What are we going to do with this lot . . . the punters?'

'You could chair it yourself,' Keir said.

Annie shook her head. 'It's not my subject.' They were going to debate the Falklands – had the war been avoidable or had it not.

'I could brief you,' Keir said. 'I did Jeremy's brief, and it's all in my head.' Hope was hammering against his ribs but his head was icy cool. It didn't happen like this, a fairy-tale chance to step into the spotlight. But Annie was looking at him, alternately biting the inside of her cheek and her lower lip.

'Could you handle it? We've got nothing to lose, and we could see if there's the germ of a programme. We never intended to put the pilot out, anyway, and at least I'd have something to show for the outlay.'

Keir tried to look as though he were considering. 'I think I could,' he said. He knew he would be utterly bloody brilliant but he couldn't afford arrogance, not yet.

'Go to make-up . . . you'll need a jacket. Where's Wardrobe? And Keir . . .' He looked at her, his mouth dry, but all she said was 'Good luck.'

He walked off the set slowly, only accelerating when he was outside the lights and the need to punch the air could no longer be contained.

They were saying something about bombs in London on the radio but Kath wasn't paying attention. Pain had gripped her as she stood at the sink, thrilling through her as though she was a lightning-rod. When it eased she switched off the radio and reached for a tea-towel to wipe her face. That was better! She looked at the clock: Dave would be down the pit now, no point in summoning him back. Could she sit it out till he came home? She had never wanted anything so much in her life, but this was a second baby – they did come more quickly, and Brian had scarcely been a difficult labour. She would have to turn to her mother.

There was time for a decent cup of tea, though, the last she would get for a few days. Hospital tea was vile. She sat by the fire to drink, hand on a belly gone strangely still now after the activity of the last few months. Would it be a boy or a girl? It didn't matter, really . . . except that she already had a son. A daughter would be nice for Dave . . . and mothers and daughters were always close, or so they said.

Kath thought of her mother's implacable expression, to be faced in an hour or so. She would take Brian in, no doubt about that, and she would offer good wishes for a short labour. But she would do it as though for a stranger, the girl up the street. And they had always been so close; Kath would have sworn nothing could have come between them. 'I've disappointed her,' she thought and was suddenly heartsick.

When the pains were coming at twenty-minute intervals she rang the hospital to say she would be arriving and scribbled a note for Dave. With a bit of luck it would all be over by the time he got home. She made a final call, to a taxi, and shortly afterwards bundled herself and Brian, each complete with a tartan case, into the cab.

It was only two hundred yards to her mother's house. 'Wait here,' she told the cabbie and struggled out, the baby on one arm, his bag on the other.

'Need a hand?' the cabbie said but he didn't move and his relief when she turned him down was almost palpable.

Kath lifted a hand to her mother's door, and then changed her mind. This was her home, or it had been – surely she had some rights left? There was the same old clutter in the yard, the same green paint on the door, the smear of white paint on the brickwork where Michael had

thrown a loaded brush at Tommy. Only the welcome
would be different from the old days. She pushed the back
door open and stepped inside. Her mother was standing at
the sideboard, watering a straggling Busy Lizzy from a
milk-bottle.

'It's time, mam.' She had hoped for a smile, a hug, some
sign of softening but none came.

'Off you go, then,' her mother said, taking Brian into her
arms. 'Get it over and done with.' She might have been
speaking of a tetanus injection.

Kath's youngest brother, bare-footed and sheepish,
appeared behind his mother as Brian looked curiously into
his grandmother's face.

'Want company, Kath? I'll come with you. Just till Dave
gets there, like.'

Kath felt her eyes prick. 'Thanks, Michael, I appreciate
it. But I'll be all right. I don't want all the nurses flocking
round me handsome brother, I want them concentrating
on me.' Did her mother's face, watching at the window
with Brian, soften as the cab drew away? A pain gripped
her, more intense than the ones before, and she sat back,
legs together, hoping she could hang on. In the mirror the
cabbie's face was a picture of horror and curiosity.

'Don't worry,' she told his reflection, 'I think I can wait
till we get there.'

'It's a good job, missus,' he said, revving the engine like
a demented James Hunt, "cos I faint if I cut me finger.'

Jenny was editing a tape from an interview at the Barbican
when word came that Bart was down in the lobby. She
resisted the temptation to check on her appearance and
went down in the lift, firming her resolve as she went.

He was standing by the reception desk, his briefcase
propped between his feet, a paper-wrapped sheaf of
flowers in his hand.

'Pax?' he said, wrinkling his forehead as though expecting a blow. 'I've been up to my eyes, Jen, on the Freddie
Laker thing, but it's done now, and I'm back. If you'll have
me?'

She would have him all right, but not without at least token resistance.

'I'm sorry, Bart, but I'm in editing and someone else is waiting to use the suite, so I'll have to be quick.'

'Lunch?' he said, holding out the flowers.

'Freesias . . . lovely. Sorry, lunch is out. I could get away about half-five.' Suddenly she decided to be truthful. 'I'm meeting someone else for lunch: Euan Craxford.'

Obviously Bart didn't regard Euan as a threat. 'That's nice. Have a good lunch, then, and I'll see you at six. The Egypt?' It was her favourite pub and he knew it, the swine.

'OK, as near to six as I can manage.'

Jenny finished the edit and carried the tape back to her office.

'Any good?' Simon shared her office and her job on the arts magazine programme.

'Not bad,' she said. 'I've used six minutes five, so we ought to be OK.'

'Coming down the pub?'

'No, thanks, I'm meeting someone for lunch.' He was eyeing the flowers she was arranging in a plastic vase.

'Bart again?'

'What d'you mean "again". It always has been Bart . . . well for ages, anyway.'

'Did I say otherwise? Don't flash fire at me, Jenny, you know my heart's weak. It's just that freesias seem to be his calling-card after he's had a lost weekend or so.'

Jenny liked Simon but there were limits. 'Thanks very much, Claire Rayner. When you've disentangled yourself from your many liaisons I might take your advice. Until then, keep it to yourself.'

But she couldn't help pondering Simon's words as she waited for the lift. *'Bart's calling-card'*. Bloody cheek! They were almost down to the ground and she was preparing to greet Euan when someone got in at first-floor level.

'Have you heard? There's been a car bomb. The IRA. Two guardsmen dead and dozens of injuries. It was the Blues and Royals and they say the road's littered with dead horses. God, they're even taking it out on dumb animals now.'

Jenny had been looking forward to her lunch with Euan but what with Simon's barbed words and the IRA bomb her appetite was dwindling. Only the sight of Euan, eager and familiar, persuaded her to smile and take the proffered arm. His red hair was neatly brushed for a change, only threatening to curl here and there, and he seemed taller and more elegant in his well-cut three piece suit.

'It's lovely to see you, Euan,' she said and held up her face to be kissed.

A second bomb had exploded, this time under the bandstand in Regent's Park. Six dead and a score or more of wounded.

'The Green Jackets,' Steven said. 'It was the Blues and Royals in Hyde Park, and now it's the Green Jackets.'

'And a bandstand,' Cissie said, shocked. 'There might've been women and children there.'

When Steven switched off the TV he tried to read the paper, but the trivia it contained were more than usually irritating in the presence of tragedy. He put it down and turned instead to his favourite topic. 'That phone's a waste of money.'

'So you've said.'

'So I'm saying it again. We never use it. It never rings us.'

'Well, it hardly can, Steven,' Cissie said reasonably, adding to his chagrin. 'I mean, it can't make itself ring.'

'I know, I know. Your fancy son never rings, that's one thing I do know.'

'If you mean Keir, he rang last week to tell us about him and Toni breaking up. And while we're on the subject, why "my" son? How come he was your son when he was winning scholarships, but when he doesn't dance to your tune he's all of a sudden "my" son?'

If he'd had the energy he'd've got up and gone for a walk, but it was hot and he couldn't be bothered. He picked up the *Mirror* again and shook it out in front of his face to indicate that relations were strained.

'Want a cup of tea, chuck?' She was suddenly there, one

floury hand at his temple, the other manipulating the pages of the paper to find her stars. *'Jupiter in Pisces means you will find relationships difficult,'* she read aloud. 'Well, that applies any day in this house.' Her hand on his ear was tickling pleasantly and he would have liked to turn his face against her pinny and make friends. Nothing to stop them mounting the stairs together. But she was reading on: *'A small windfall could be a pleasant surprise today, and there may be news from abroad.'*

Steven fished in his pocket and brought out a new one-penny piece. 'We know no one abroad, Cissie, and you don't do the Irish Sweep, so here's a small windfall. Now give me peace. Put the kettle on while you're on.'

Steven had hoped to find news of Poland in the paper but it was all kiss and tell. He would get a different paper if he had his own way, but Cissie swore by the Stars and the Letters Page. There had been martial law in Poland for nine months now, and clashes, especially in Gdansk where Solidarity had started. It would fizzle out, of course, as every move to freedom had fizzled out in the last twenty years. The Russians had a grip on Eastern Europe that would never ever be broken. He tried to remember how long other empires had lasted: hundreds of years, probably. It would be the twenty-first century before the Poles got freedom . . . and then some other bastard would rise up and take it away from them!

'Steven!' He knew that tone in Cissie's voice and it boded no good. 'Steven, I hope you're not going to sit around all day like that?'

'Like what?'

'You know, unwashed, unshaved, dirty old shirt . . . and those trousers!' She was shrugging into her good cream coat and putting on a head scarf.

'Where are you going?'

'Where I told you yesterday . . . and the day before.'

'Mebbe you did tell me but I can't remember.'

'That's because you never listen. The only time you listen to me is when I say "food". You sit up then, all right.'

'Where are you going, Cissie? Either say or don't say, I'm not begging.'

233

'I'm going to church, Steven. You know, the big grey stone place on the corner of Church Street where they marry people. You went once, thirty-odd years ago. They're having a coffee morning there, and I want to see if Kath Botcherby's bairn's come. She's well due.'

'Big event!' Steven said scornfully. Cissie was stung.

'Every bairn's a big event, Steven Lockyer. Apart from which there'll be a christening, won't there? All our Keir's friends'll be there, and I don't want to miss it.'

'They won't ask our Keir?'

'I expect they will ask him. But he can't run up and down the country like a yo-yo, he's got a job to consider.'

Cissie would make excuses for the lad and perhaps it was just as well that she could. But Steven could face the unpleasant fact that their son did not come home for one simple reason: that he didn't care enough to bother.

'We can go on to Elaine's place later,' Euan said as he and Alan settled at a table and accepted huge *à la carte* menus. 'I lunched with Jenny today, and she sends love.'

'Is she well?' Alan asked.

'Blooming,' Euan replied. 'And dinner's on me – I'll tell you why later. What do you want? It's by way of being a celebration, so dive in.'

After five minutes of vacillation Euan took the reins. 'Smoked salmon, *truite aux amandes* and *filet de boeuf* Diane.' And then to Alan, 'Is that all right?'

Alan raised his eyebrows. 'Perfect,' he said, his lips twitching.

'And a bottle of Bourgogne Aligoté to start, please,' Euan said firmly. 'I'll think about what comes next.'

As the wine waiter withdrew Alan said: 'You've come into money?'

'Cor-rect,' Euan said. 'A very decent old uncle of my mother has departed this life and left me his all, on the grounds that as I'll never provide for myself he'd better do it for me. I wasn't flattered. Pleased . . . but not flattered.'

'Count your blessings,' Alan said, looking round the restaurant. 'Are you taking advice about investment?'

'That's all been done for me,' Euan said and Alan looked

at him sharply. Euan caught the glance and pulled a face. 'They're doling it out to me, six months at a time. My mother tells me that at present rates I can live on the interest, which is very nice except that it's back to the status quo: I've always had an allowance, and I shall still have an allowance albeit much larger and from a different source. So if the old boy meant to give me freedom it failed. He was a decent old stick, too. Moved in with the local barmaid at one stage. Did him the world of good.'

'At least you've got security,' Alan said. 'Most people would envy you.'

'I know, and I do appreciate my luck in some ways. But I used to envy you all, especially Jed, because you were all the masters of your fate. You might sink, you might swim, but whatever happened it would be of your own volition. Does that make sense?'

Alan waited until the first course had been set before them and the wine tasted and approved. 'In theory it makes sense, but no one is really free, Euan. We're all enslaved to one degree or another.'

'Not you, surely? Certainly not now.'

'I've never been free,' Alan said. 'There's never been a moment that I wasn't aware of the burden of my mother's expectations. I won't go in to it now, but it hung in the air at home. And now, when there's no one to answer to, it's worse. I feel more indebted, more bound. It never was a spoken thing, no one said to me "You must succeed." But it was there just the same, and it still is.'

Euan nodded sympathetically. 'I know what you mean . . . but I'm right about Jed, aren't I?'

'Yes, I think Jed really was his own man underneath all the left-wing spouting. Jenny is free, too: there's parental hope there, but they keep it in bounds. Barbara is hagridden, but of her own volition. Who else? Keir . . . Keir was more tied than even I, but he's broken free.'

'His father's a socialist, isn't he?' Euan said. 'Up the workers and *viva* Scargill?'

'Have you met him?' Alan asked. 'He's a nice man but he feels he has to be bitter about society – so you see he's enslaved, too. We all are.'

'You've missed out Kath and Elaine,' Euan said. 'Kath's

mother is still being abysmal to her, and poor Elaine is barely able to keep L.J.'s head above water. I mean to change all that.'

'Ah ha,' Alan said, raising his glass. 'Do I get out the morning dress?'

'I've got plans,' Euan said grinning. 'Musical plans. If they come to fruition Elaine will be all right. She won't take money from me, not even for L.J. – it maddens me, but I respect her for it. And talking of morning dress . . . how's your love-life? I used to work it all out when I was playing at farming. I'd kill a few flies and then partner you with Jenny, Keir with Kath . . .'

'You got that one wrong. She can't stand him, never could.'

'In any case she spoiled it all by marrying out of the club. Silly sausage.'

'I haven't seen much of Jenny since my mother died. She was so kind then I felt I mustn't play on her good nature. There's a man now, isn't there? A journalist.' He raised interrogatory eyebrows.

Euan picked up the bottle and refilled their glasses. 'He's what I call a Fell-type.'

Alan looked uncomprehending and Euan put down the bottle. 'I do not like thee, Dr Fell, the reason why I cannot tell. But this I know and know full well, I do not like thee, Dr Fell.' He finished the rhyme and lifted his glass. 'To the legal profession, wily bastards that they are. You neatly sidestepped my question about your love-life. You're on oath now, so give me the facts.'

'There is someone,' Alan said, as the waiter reappeared at the table carrying their next course. 'In my chambers . . . but it's early days. And that's all you'll get, so let's get on with the food.'

Keir carried the wine into the bedroom and set it on the marble-topped vanity unit. Annie was moving about the room, casting off garments as she went. He knew the routine now. In a moment, when she was naked, she would circle the room again, bending to gather each discarded item and carry it to the linen basket. He began to unbutton his shirt, watching the muscles ripple in her lean body with its faint tide-lines of tan at breast and thigh. Annie Love was at least forty, probably forty-five. He didn't find her age off-putting; if anything, it excited him. After they had made love she would lie, arms above her head, eyes closed, seeming oblivious of his interest, while he explored her body with eyes and sometimes fingers, seeking traces of ageing, diminished elasticity, the blueing of veins. But however hard he looked he could find nothing except a faint reddening, a minor crepe-ing of her neck, a single white hair here and there among the clusters on her body.

She had finished collecting her soiled clothes, and he too was naked. They moved together to the shower and stood, gasping, under the flow while they soaped and lathered whatever came easily to hand, not caring whose body it was so well did they fit together.

While they wrapped themselves in the thick blue towels Keir tried to work out if he would have fucked her anyway, without the added bonus of possible advancement. 'Possibly,' he thought. 'Probably,' he decided as he followed her into the bedroom, letting the towel fall as he climbed on to the bed.

'God, I'm bushed,' Annie said, turning on to her face. He put his hands on her shoulders and began to massage her, feeling a rising excitement. Sometimes, when she was in this mood, she would talk about work – bitching mostly, and getting rid of the angst of the day, but sometimes formulating plans. If he got it right, pleasuring her enough but not so much as to make her turn and reach for him, she might tell him what they thought of the pilot programme and whether or not she had meant it when she had praised him as the cameras ceased to roll.

Suddenly, in his head, he heard Jed's voice: 'You're a cynical sod, Keir Lockyer, but you'll go far.' Jed had said that more than once. He hadn't thought of Jed for ages. Funny the way you could forget things. He hadn't rung home either. In a way it had been better when they didn't have the phone and it was an impossibility.

'Keir, for Christ's sake do it or leave it alone.'

'Sorry,' Keir said, kneading the strong brown shoulders, and tried to get back the rhythm that produced results.

'One more . . . one more.' There was no rhythm to Kath's pushing now. She wanted it out of her, no matter what.

'Oh God,' she gasped, the words turning into a wail . . . and then she felt her body give way and the baby was out in the world.

'Clever girl, clever girl.' The midwife was half-laughing, half-crying. She was quite old, fifty at least, so this must be her umpteenth baby, but she was still excited. 'It's a girl . . . and listen to those lungs. You wanted a girl, isn't that nice? Lie still, darling, you can have her in a moment.' She could hear them working on the baby, sucking out the tiny mouth, weighing, measuring.

'She's a big girl.'

'I could tell that,' Kath said, with feeling.

'Got a name ready for her?' the pupil-midwife asked as she sponged Kath's hands and face and got her into a clean nightdress.

'*Deo gratias*,' Kath said solemnly, still panting.

'Oh, nice,' the nurse said, taken aback.

'It means "thank God",' Kath said. 'Thank God it's over.'

'Oh, I see,' the nurse said, relieved.

'We'll probably call her Margaret – it's my mother's name and my mother-in-law's name, so we'll have to use it somewhere.'

'She's *lovely*, Kath,' Dave said, when he came in, self-consciously carrying flowers. He was grinning foolishly at the sleeping baby.

'What've you done to your hand?' Kath said, seeing the stark white bandage on his index finger.

'It's nothing . . . just the tip. It got caught on the conveyor. Don't panic, I can still use it. See!' He wriggled the huge finger to prove his point.

Kath took his hand tenderly, holding it against her cheek, seeing the other scars of other battles with the pit. 'I can't bear you to get hurt.'

'That goes two ways. No more bairns, Kath – one of us is having the snip. Two's enough to provide for, anyway, but I don't like seeing you suffer. The last month's been a nightmare.'

'It's worth it, Dave. Being pregnant is hell on earth, I won't deny it – that's why women go through labour without too much fuss, they're so damn glad it's over. But then you see the baby. I'll tell you something, Davey Gates, if the Princess of Wales is half as happy with her bairn as I am at this moment, I'll be surprised.'

Around them the ward hummed and chattered, but all Kath could see was Dave's face, shining now with pride and sentiment and embarrassment.

'Pretty,' he said, touching the frills of the pink nightgown she had bought for her confinement. She lay back on her pillow, feeling suddenly tired and overcome with love for him and the new life in the crib. She looked right, seeing another Dave bending to his wife, and then left where a third man was paying homage to the crib. Each birth was a miracle and it always would be, to the end of time.

She put her hands on the mattress and heaved herself up, feeling her body bleed a little at the movement. She must have winced, for Dave's face contorted in sympathy.

'OK?' he asked.

'I feel wonderful,' she said, and meant it.

22

15 August 1982

Jenny had forgotten how nice it was to be at home. In the train coming north she had wished herself back in London, had even counted the hours to her return. Then the Cleveland Hills had come into view, familiar and reassuring, and after that Durham, the cathedral towering into the sky, the castle squat and protective below it, and she had been suddenly proud and glad to be back.

Now she lay in her own bed in her own room, seemingly untouched since school days, drinking in breakfast smells from down below, listening to the morning service faint from the kitchen radio, revelling in being cared for once more.

'You look peaky,' her mother had said when she arrived and appeared to accept the excuses Jenny made for Bart's non-appearance. She couldn't possibly have told them the truth, that he had said: 'It's not my kind of thing, Jen. I've never been to a christening, if I'm honest.' She might have accepted that if it had not been for an uneasy feeling that his refusal to come north had a lot more to do with meeting her parents than attending the christening of her best friend's daughter.

'There now,' her mother said, bustling into the room with a tray and a newspaper, waving aside Jenny's protestations, 'it's a long time since you were spoiled, I bet . . . living on your own. I've done all your favourites. Eat up. You've got plenty of time and daddy will be ready with your lift whenever you say the word.'

When Jenny had eaten and put the paper aside she lay back and thought about the coming ceremony. She was to be a godmother, to hold Kath's baby in her arms and make promises. Was she capable of living up to those promises?

She would care for the child materially if ever Kath and Dave could not, but the spiritual side was something else. She thought of the guest-room next door, prepared for Bart's aborted visit. If they knew that she and Bart slept together what would her parents think? She imagined hurt in their eyes, the pursed lips. Or was it all a colossal bluff? Had her parents indulged in free love once, and only donned their moral stance when they had the responsibility of a child's upbringing? Did they know she was no longer a virgin and accept it, as long as they were not forced to acknowledge it openly?

Thoughts of lost virginity brought thoughts of Keir. Kath had invited him as a gesture to the old days but no one expected him to turn up.

'What's he doing now?' Kath had asked the last time they talked on the phone, and Jenny had mentioned his affair with Annie Love, which was a juicy topic of conversation whenever media people met.

'Who is she?' Kath had asked curiously and Jenny had muttered something about her being a power in television circles, and had turned to news of L.J. and her love-affair with a besotted Euan.

'Why do I want to see Keir again?' Jenny wondered as she bathed. She had still not forgiven him his casual treatment of Elaine and nothing she had heard of him in the last two years had been to his credit. 'He hasn't had an original idea in his life,' one researcher had told her, 'but he's pinched every good idea he's ever come across and used it as his own, hence his reputation for fizz. I wouldn't work with him again.'

Had the girl been unfair? Keir was clever, brilliant even, according to Alan. Why should he poach other people's inspiration? Unless it was because he couldn't wait for success. She had seen that in some of her own colleagues, a desperate impatience to be recognized, acknowledged.

She pulled out the plug and reached for her towel. Time to put on her new beige suit and her Dior ear-rings and tell the world that Bart's absence didn't matter a fig.

They sang 'Morning has broken' and 'All things bright and beautiful', and then the babies were being carried forward and the respective families were crowding round

241

the godparents and the font, and the vicar was taking each child in turn to be named and dedicated to God. It was a plain church, sparsely adorned, but to Jenny's eyes it was beautiful, filled as it was with family worship.

She carried the newly named Margaret Emma out into the sunshine, Margaret for her grandmothers and Emma after Jane Austen's heroine, whom Kath had studied in depth for her English A-level. 'It sticks, Jen, it sticks. You think you've forgotten it all and then it comes back to remind you. And the first person to call her Em will get thumped.'

Back at the tiny, terraced house with its bullseye-glass front door, the baby was given to her father, and Jenny and Kath repaired to the kitchen to uncover the sausage rolls and pies and bowls of trifle.

'God, you must've worked like a slave,' Jenny said, putting milk into a trayful of cups.

'Dave's mam did most of it. She's got a lovely pastry hand.' Kath sounded wistful and Jenny laughed aloud.

'What's funny?'

'You,' Jenny said.

'Thanks a bunch.'

'No, it's a compliment really. It's just that you're so . . . domesticated. You said "pastry hand" with so much reverence.'

'I meant it,' Kath said ruefully. 'I'm no great shakes as a housekeeper. But when I apologize Dave says that isn't what he married me for.'

'I should think not. You're a good wife and mother, that's what counts. I was only laughing because, of all of us, you seemed the least likely to settle down – or that's what I'd've said a few years ago. Barbara only wanted to be Mrs Something, Elaine seemed to need a man so much, I was very conventional. But you were the free one, independent-minded . . .'

'And now I'm the one who's chained,' Kath said slowly.

'You don't see it like that, do you?'

'Sometimes, if I'm truthful. Sometimes I wouldn't change places with Lady Di, palace and all, but other times I remember the ambitions I had . . . teaching in Germany. Do you remember when I was going to apply to BFO? And now you and Elaine are in London, and ye gods, even Barbara's made it to Jersey. But I'll be in Belgate for the rest of my life.'

'You don't know that.'

'I do, Jenny. Dave would never, ever leave Belgate. He wouldn't even move to Sunderland, never mind anywhere exotic. Me mam was right about that, I've got to admit. Which reminds me, she's sitting one side of the fire clamming for a cup of tea and Dave's mother's sitting on the other. If we don't get in there with the food there won't be a word exchanged.'

Keir was surprised at the extent of Annie's pleasure at the gift. He had intended that she should be pleased, but this was over the top.

'You darling! It's lovely. I can't let you give it up, though – it's a family possession.'

'I want you to have it.' Keir had bought the small gold locket in Camden Passage, choosing it carefully so that it would be good enough to please her, not so good that she would feel bound to refuse it. 'It was my grandmother's,' he'd said when he gave it to her. 'It was left to me and I want you to have it.' He had shrugged. 'I know it's not much but . . .'

She had been dressed to go out to lunch when he handed over the gift, and now she rushed into the bedroom to change into a dress that would provide a more suitable setting for the locket. It was strange to see her so child-like in her enthusiasm. Usually she was in total control of her emotions, but today he had managed to touch a nerve. That meant that he was beginning to understand how she ticked. He felt a glow of satisfaction and turned to the mirror to check his cheek for lipstick.

There was the faint murmur of conversation from the bedroom: Annie must be on the phone. He considered lifting the other receiver and eavesdropping, something he had done before, but it was Sunday. She wouldn't be talking shop on a Sunday, so it was hardly worth the agonizing effort of letting the receiver-rest rise too slowly to give a tell-tale click.

When Annie came out of the bedroom she had put on a peacock-blue sleeveless dress with a low neck. The locket hung at her reddened throat, above the faintly freckled breasts.

'See,' she said, 'it looks wonderful.' She was moving to kiss him again and Keir tried to let eagerness show in his face. She stopped short and looked him in the eye.

'I've got a present for you, too. I gave the cassette of the pilot to Jacob Elliman on Friday, after I'd edited it. I've just rung him, and he likes it. He says we can do a one-off, and see how it goes.'

Keir tried to feign ignorance. 'Good. That's good! Will Jeremy be OK by then?'

'Not with Jeremy, idiot! With you. He says you're gauche, cheeky, and he loves the Geordie accent, so it's worth a tumble, according to him. And now let's go and eat. I'm starving.'

'My God,' Keir said, feeling real amazement contort his face. He had never believed it, not *really* believed it. And now it had come true. 'You bloody little darling!' He seized her waist, lifting her until her feet left the floor and they were face to face.

'All right,' she said, 'don't explode.' But he could see she was pleased by his delight.

'God, Annie, I'll make this up to you.' He lowered her to the floor and put his hands on her shoulders. She was smiling, looking strangely vulnerable, not like the old tough Annie Love. He moved his hands over her shoulders, seeing the line, like a garotte, at the base of her neck, the slackening of the skin beneath her chin. Suddenly he loved the signs of age in her, wanted to cherish the older body, lend it his own youth.

'Sod going out,' he said and lifted her in his arms to carry her to bed.

'I'll do the pots,' Dave said, from the depths of the armchair where he had sunk on departure of the guests. He made no move and Kath hurled a cushion at him.

'Bloody fraud! If you went out to that kitchen and put your hands in suds they'd drop off.' She turned to Jenny.

244

'This is all for your benefit, Jen, so you won't go back to London and say I'm married to an MCP.'

'A what? I vote Labour,' Dave said.

'Funny, funny. You know what MCP means because I've told you. Still, we'll wash up and you can bath our Brian and put him to bed.'

'After which you are both going out,' Jenny said firmly. 'I'm baby-sitting. You don't often get out together and tonight should be a celebration.'

'Are you sure?' Kath asked, eyes anxious.

'Very sure,' Jenny said. 'I'm out most nights in London, so a night in will be a treat. What's on telly?'

'Hold on,' Dave said. 'I like the way you two are carving my night up. I'm not sure I can take Kath down the club, I don't know what the lads would say. They might think I was hen-pecked.'

'You'll be more than hen-pecked if you don't watch out,' Kath said, moving towards the kitchen door.

'Come here, woman,' Dave said, catching her by the wrist and pulling her down on to his knee. She made a show of protest but Jenny could see how she responded to her husband.

'They love one another,' she thought and the idea both pleased and saddened her.

'Will you marry Bart?' Kath asked later, up in the bedroom where Jenny was arranging Kath's hair for the night out.

'It's not that simple,' Jenny said, back-combing another strand. 'I don't know if he'd ever ask me, I don't know if he's the marrying kind. He likes his job . . . it's important to him. He might think marriage would tie him down.'

'Well, it would,' Kath said. 'But most men accept that eventually. And it's not so bad. We had hard times at first, getting a home together on one wage, but the good times are coming. Dave's a deputy now, so things are better. And I'll work eventually – I can't let all that slog at college go to waste. We'll be in clover in three years' time. 1985: that'll be our year. You and Bart are high-earners anyway. And you could work if you had a kid – get a nanny?'

'On BBC pay? You must be joking!'

When she had waved Kath and Dave off Jenny checked

on the sleeping children. They lay on patterned flannelette sheets, rosy and spotless, small mouths open in contentment. Would Bart like a child? What if she let herself get pregnant and presented him with a *fait accompli*? Would he knuckle down to fatherhood or simply walk away? *'It's not my thing, Jen, christenings. I've never been to one, I don't want one of my own.'* That's what he would say and she would be left to cope like Elaine, doing justice to neither job nor child.

She went downstairs to the tiny sitting-room, where the fire burned bright to heat the water in spite of the warm August air outside the open windows, and the arms of the easy chairs were worn into holes. Perhaps Kath was the lucky one, after all? Not a drop-out but a victor? In three years' time she would have it all. 'What will I have then?' Jenny thought and was suddenly afraid.

BOOK THREE
1985

23

7 January 1985

Kath got down on her knees and began to mend the fire, using cinders and bits of stick, twisting newspaper into tubes and knotting them . . . anything to substitute for coal. Dave had dug half a bucket of slack from the railway embankment, and when that was gone she would have nothing with which to keep her children warm. For the first time since the strike had started they would be without a fire.

Britain's miners had been on strike for almost eleven months, and bitterness pervaded the Durham coalfield. Now it was even affecting the children. Kath felt tears prick her eyes and brushed them angrily away. Things were too bad for crying.

She took a sheet of newspaper and tried to coax a blaze by holding it either side of the fireplace. She was waiting, hoping to see a sudden glow, even to see the paper blacken and burst into flame, when she suddenly noticed Keir Lockyer's face smiling out at her. *'Keir does it again. TV star to host BAFTA awards ceremony.'* So he was going to be on telly once more! According to Dave that was one good reason for letting the telly go, as they had also had to give up the telephone and the car. They had sold everything that wasn't rented or on HP: soon there would be nothing left except the house. 'And then I can stop worrying,' she thought. 'There'll be nothing left to go then.'

She tried to concentrate on Keir's picture again. The hair was blonder and curled slightly; it was permed but that was only obvious to someone who had always known him. The skin was dark, bronzed with a sun-lamp presumably since he never took time off to holiday.

Kath thought back to the old days, to the dawn on the

hill when they had all bragged about the future. How cocky they had been then. Jed had said something about them being the future, and three weeks later . . . no, two . . . he had been dead. At least she was alive. And she must blaze up this fire: she couldn't keep the bairns in bed much longer and if they came down now they would freeze.

The door creaked open and Dave came into the kitchen, white-faced and shivering but smiling just the same. 'There you are, Kath. Logs. That'll get a blaze.' She tried to look pleased at the sight of the green wood, saplings cut down from every wooded place around Belgate, but all she could think of was the tree that might have been. Each sapling filled one grate for one day. Left to grow, those same saplings would have given pleasure for a lifetime.

'That's lovely, pet,' she said aloud and got wearily to her feet to make him tea.

While she waited for the kettle Kath wondered how much longer the strike could last. Every day some of her neighbours were in court for picket offences or stealing coal. Some men had snapped after ten months and had gone back to work at Christmas: now they were pariahs. Belgate was a divided village; the only thing they all had in common was misery. How could a strike which the majority did not want have dragged on for so long?

There was a knock at the back door and Kath went to open it. Her mother stood on the step, her coat clutched around her, her head shrouded in a woollen scarf. She held out an old margarine carton with two eggs in it.

'For the bairns,' she said. 'One each.' She was already turning away as Kath spoke.

'Are you all right? Can you spare them?'

'We're managing.' Her mother said nothing more but her eyes spoke volumes before she turned away. 'See,' they said. 'I told you he'd bring you to this.'

Kathy stood for a moment, watching her mother's retreating figure, wanting to call out and bring her back but unable to find the words. At last she shut the door and went back to her kitchen.

Outside in the street she could hear children chanting on their way to school. Not a nursery rhyme or a pop song –

the tune was 'What shall we do with the drunken sailor?'.
The words were new.

What shall we do with Maggie Thatcher?
What shall we do with Maggie Thatcher?
What shall we do with Maggie Thatcher early in the
  morning?

Burn, burn, burn the bastard,
Burn, burn, burn the bastard,
Burn, burn, burn the bastard, early in the morning.

'We've got Kim Howell for the union, Michael Mates for
the Government and Paul Kovalski for the Opposition,'
Jenny said into the phone. 'I'm hoping to get Paul Johnson
as fourth voice, but he hasn't given a firm yes or no yet. If
I can't clinch him by eleven I'll go for Ann Leslie.'
  At the other end of the line her boss grunted approval
and made suggestions.
  'Don't forget the strike's been done to death, Jenny. It's
been going on for nine months, after all.'
  'Eleven,' Jenny corrected. No one who cared about
Durham could possibly forget how long the strike had
lasted.
  'OK, eleven . . . which makes my point. Tell Sue to go
for new angles, not the hoary old issues. And be careful
about the arson case. *Sub judice*, unless it's settled today.'
Nine striking miners were on trial for arson and their fate
would probably be decided that afternoon. Jenny glanced
out of the window and saw flakes of snow falling from a
leaden sky. God, this was a winter of trial, and no mistake!
  When she had put down the phone she pulled out a
sheet of paper and wrote a quick note to Kath. They had
talked often on the phone when Kath still had it, but now
there were only letters. When Jenny could think of an
excuse she sent money 'for the children', but she knew that
an unexpected letter sometimes revived flagging spirits.

All the vibes down here are that it will be over soon,
Kath. Hang in there. I'll be up for a weekend soon. Now

251

that Bart and I are defunct I am a free agent so watch this space. Elaine and L.J. send love. She's still battling on but L.J. is getting to be quite a handful. Delightful as ever but a little madam. Kiss Brian and Mags for me and tell Dave I'm doing my bit for the cause down here.

Jenny collected coffee from the machine in the corridor and settled back at her desk. The strike was the topic for this week; the Kim Cotton surrogate baby for next week if the High Court had ruled. She pencilled in Clive Sinclair for the week after that: he was going to unveil his new invention in the near future. If what she'd heard from the motoring correspondent was true the C5 battery tricycle was a bit of a joke, but after two heavy weeks something light-hearted might be a good idea. When Jenny had been given a discussion programme to produce, her brief had been to inform and entertain, but with everyone who appeared on it trying to grind an axe that was not always easy.

She looked at her watch. Time for an early sandwich lunch. Tonight, after the programme was safely recorded, she was going to Elaine's to eat Chinese, courtesy of Euan.

'I will eat and eat and we'll reminisce and it'll be lovely,' she told herself firmly. 'And I won't miss Bart at all.' But in her heart she knew that however demanding and satisfying her job, however much she enjoyed her new flat in Kensington, something was missing in her life.

'I'm off then, Steve.' Cissie was tying her head scarf as she spoke and he felt a sudden anger at the eager expression on her face. Since they had opened the soup kitchen in the church hall, Cissie had been like a dog with two tails.

'Aye, get along out of it, don't worry about me. Get down there and see them fed. I'm only your man, I don't count.'

Women were lucky, really: they never retired. He had gone out seeking wood or duff coal since the strike started, but other than that what could he do? Once the strike was over he'd be back to watching telly or gazing into the fire. He had taken his redundancy a year ago, more or less forced into it, and he was no nearer getting used to it than he had been on the first day.

'I'll be back early,' Cissie said, as she left the kitchen. 'Our Keir's on this afternoon, mind on. On that quiz programme.' She kept a list of Keir's appearances, culled from the papers or his infrequent letters or phone calls.

More fool her, Steven thought, when she was gone. She persisted in behaving as though the lad still cared when any fool could see he'd tidied them away, and for good. It embarrassed him to be asked about Keir these days. They pulled his leg at the Labour Club, telling him to stand a round now they were all millionaires. 'I bet he sees you and Cissie all right,' they would say. Or, 'How's your shares in David Frost, then?' Keir was often called Durham's answer to David Frost, but their Keir had had twice Frostie's brains if he'd only used them for something sensible. This afternoon he would be on there in some pansy jacket smirking like a lecher in a brothel. It wasn't a man's job.

Discontent got Steven out of his chair and as far as the door, but the leaden sky drove him back to curse the flickering fire with its maximum smoke and minimal heat. In desperation he flung on the last of the coal he had dredged from the beach. It spat and spluttered, an occasional stone cracking and jumping out so that he had to put the guard up and cut off the heat altogether.

'To hell with this,' he thought and felt in his pocket to see if he had enough for a pint. He put on his muffler, folding it across his chest under his jacket, and pulled his cap well down on his head. If anyone had told him this was what life would be like at sixty-two he'd have jumped down the shaft long ago, while he had the chance.

He wound up in the soup kitchen, slipping into a seat and putting his folded cap in his pocket. The women had served stovies and broth to more than a hundred miners and their families, and the mound of dishes now waited to be washed and put away ready for tomorrow. He could see Cissie chatting to Kath Botcherby. Behind them other women were checking stocks, planning menus, trying to break the monotony of soup and potatoes, potatoes and soup.

A woman appeared in the doorway, a cardboard box in her arms.

'Hey up lasses, corn in Egypt here. I went down the Paki shop and told him we were desperate. He's put all sorts in. Corned beef, beans, a packet of curry powder . . . ho, ho, ho, what about the poppadums? . . . two jellies and a big tin of pears. Bless his little cotton socks.'

'Remind me to shop there when the strike's over,' a woman said. 'I'd go round now, but I don't expect he takes washers.'

'We're scraping the barrel,' Cissie said when she and Kath came to sit with Steven. 'It's all very well trying to milk the shops, but some of them must be feeling the pinch too, now. There hasn't been a decent wage-packet in Belgate for nigh on a twelve-month. There's only the likes of me and Steve, pensioners, still got their weekly money, and we're not big spenders.'

'It can't go on much longer,' Kath said stoically.

'I seem to have heard that before,' Cissie said drily. 'It's you young'uns I feel sorry for, the ones that's still making their way. How do you manage with bairns? You can't tell a bairn to pull its belt in.'

'Not very well,' Kath admitted.

'You've got a degree like our Keir, or some such,' Steven said. 'Haven't you thought of going out to work yourself, just to tide things over?'

'I've thought of very little else, Steve. But I'm married to a Durham man, remember, and he doesn't take kindly to the idea of a working wife at the best of times. Me going out while he sits at home and watches the kids is very definitely not on his agenda.'

'Mebbe not,' Cissie said, 'but he can't expect you to watch your bairns go hungry either, can he? It's men who make strikes, it's women who pay for them.'

'Would that have washed with your Steven?' Kath said, raising an eyebrow first to Cissie and then to Steven.

'Not for a minute,' Cissie said cheerfully. 'He'd've had my eyeballs for ear-rings. Tell you what, I've got a few jars of rhubarb jam put by. I'll pop you one round tonight, and a nice flat cake. Women like us with . . . what is it? chau . . . chau . . .'

254

'Chauvinist pigs?' Kath said, grinning at Steven's discomfiture.

'Yes, chauvinist pigs for husbands . . . we'd better stick together.'

Keir ordered smoked salmon and a half-bottle of Chablis to start, followed by beef and a bottle of St Emilion. Annie merely nodded to show she would have the same and folded her hands in front of her, interlacing the fingers. Keir had sensed her impatience as he signed for fans when they were leaving the studio and known she was brewing up for a row. The sooner it was over the better; he was looking forward to his lunch.

'Well,' he said, when the wine was poured and the brown bread and smoked salmon placed before them, 'spit it out. Don't let's spoil the meal.'

'What do you mean?' Annie said but her eyes acknowledged his suspicions.

'You're going to give me one of your lectures. Let's get it over with.'

'If you mean I'm going to give you some advice, you're right,' she said. 'I wouldn't have called it a lecture.'

He waited, smiling tolerantly, trying not to let his irritation show.

'It's the hair, Keir – it gets shorter and blonder by the day. The tan deepens, the suits are more svelte . . . in fact the whole picture is changing. You are becoming "media man" – handsome, slick, vaguely transatlantic. Men like that are ten a penny in this business, but you stood out because you were raw. Good-looking, yes, but gauche, cheeky. You were different, Keir. For Christ's sake, why are you so determined to be the same? I've watched you lately . . . you've suddenly decided you have a good side, and you work to it. You look for camera angles. If there's a cripple or a granny in the audience you'll be up there, cooing, watching for the birdie.'

'Anything else?' Keir felt bile rise in his throat and the brown bread turn to ashes in his mouth. Bitch, bitch, bitch! He would make her pay.

'Yes, there is something else. Your researchers tell me

255

that you constantly disregard their briefs, and I'll tell you why, Keir. You no longer want to know who's bright and knowledgeable. If Chas or Shey tell you someone's worth talking to, that guarantees you'll ignore them – because if you have to talk to someone else who's bright it just might make *you* look less than sparkling. Don't bother denying it; we both know it's true.'

'Not true and unfair. But don't let that stop you.'

'I won't because I can't afford to, Keir. If you make a balls-up I suffer for it, and I have no intention of getting a name for failure because of the overweening vanity of someone I created.'

That was what he would not forgive her, Keir decided in the taxi later . . . the suggestion that he had been nothing before she picked him up. He shrugged down into his oversized cashmere coat, enjoying the feel of the fabric against his cheek. Annie had looked repulsive today. Stringy. And the way she had licked each finger while she pondered, contemplating her huge oval nails while she did it. Her huge dead nails! God, she was a crone. How had he ever brought himself to fuck her?

All the same, he would have to play a careful game, plot everything through carefully. It wouldn't be enough just to get her off his back, he would have to finish her in the business. Otherwise, sooner or later, however much he wounded her she would surely rear up and strike him.

'You know what?' Elaine said, bending to retrieve L.J.'s bricks from the floor and throw them into the big polythene tub.

'What?' Euan said, with difficulty as the four-year-old L.J. was sitting firmly astride his stomach.

'I love this flat.'

They had been in Hampstead for a year now, sharing the big airy living-room and kitchen, each with a bedroom of their own.

'You'd be doing me a favour,' Euan had said when he had offered Elaine the chance to move in with him. 'I can't live alone. You know me, I *need* company, Elaine. I'm addicted to it. If you don't take pity on me I'll be found there one day, shrivelled away for lack of companionship.'

Elaine had warned him about L.J. and the sleepless nights, her own erratic hours, her lack of domestic flair. Euan had brushed everything aside, and in the end, tempted by how much better it would be for L.J., Elaine had agreed to a trial period. She had waited for Euan to try it on, would even have welcomed it in a way: she was sick of one-night stands or give-and-take relationships where the giving was all done by her. She could accept Euan in her bed . . . in fact, it seemed a small price to pay because she really liked him and always had.

But he had never as much as made a gesture. He kissed her occasionally, the odd companionable kiss of old friends, but the rest of his affection went to L.J. Sometimes, when he likened her to Jed or prophesied great things for a child with such genes, Elaine was tempted to tell him the truth. But she always resisted the temptation. What was it Jenny had said, long ago? A secret is something one person knows. Two people knew her secret, but Jenny was like the grave, and that was the way it would stay.

'Did you hear what I said?' she asked again. 'I said I love it here.'

'So you're staying, then?' Euan asked. 'Good! I did have this belly-dancer lined up to move in but . . .'

He ducked as she flung a cushion, cradling L.J. in his arms and squealing, 'Save me, L.J., save me!'

The little girl turned on her mother, arms akimbo. 'Now stop this at once.'

'Ooh, hark at her,' Elaine said, leaning forward to tickle her daughter's bare feet.

'I said stop it,' L.J. said, mouth firming. 'Or else I'll have to put you to bed.'

'Yes, please,' Elaine said meekly. 'Then I won't have to go to horrible old work.'

'I hope you're not speaking of the Club de Naff so dear to my heart,' Euan said.

'Too bloody right,' Elaine said. 'It should be bombed, that place. All those Jap businessmen eyeing up the girls – and the watered drinks . . . it's a knocking shop!'

'The music's good though,' Euan said. 'There's this

brilliant vocalist – and the drummer! Oy veh, the drummer!'

'And the bass is a junkie and the keyboards beats his wife . . .'

'His *third* wife,' Euan finished.

'His third common-law wife,' Elaine amended. 'You know what?'

'What?'

'We're moving in the wrong company.'

'Too true, but you wait till the ship comes in. I've got a new song for you.'

'What's it called?'

'"Changes". I'll go over it with you tonight if we've got time. Let's get cracking now. You bath the little monster, I'll do her supper and clear up.'

He was half-way to the door when Elaine called to him.

'Euan! I don't deserve you.'

'No,' he said feelingly. 'A nice girl like you – you certainly don't.'

They arrived at the club to find bored girls lounging at the bar, waiting for the influx of visiting businessmen with esoteric appetites and bulging wallets, but the tables were empty as yet and canned music jangled from somewhere off-stage. The lighting was always subdued but now, before opening, it was almost non-existent, giving an air of unreality to the shoddy scene.

The keyboard player and the bass guitarist had already arrived and were scrounging for food behind the scenes. They were old hands, hack-players happy enough to churn out cover versions of old standards night after night in return for a steak supper and a couple of hundred quid a week.

'Play me the new song,' Elaine said. 'Quickly, before anyone comes.'

'I haven't got it quite right,' Euan said. 'And you know I'm no pianist. Still . . . it goes like this.'

Elaine listened, head on one side, humming a little as the melody emerged. 'I like it,' she said at last. 'The melody's strong and the words are lovely. Really memorable.'

They went off to change into their performance clothes

before doing the nightly sound check, and when Elaine came back, her dark hair swept up above the cerise lamé sheath dress, a man was sitting at the piano.

'This is Carl,' Euan said. 'Carl Smiley meet Elaine Gentry. Carl's staying around till the break, Elaine. You can talk to him properly then.'

'Who is he?' she hissed as the man walked away. 'He's a bit dishy.'

'He's part of my master-plan,' Euan said, grinning. 'Now let's get on with the show.'

Elaine moved smoothly into the set but her eyes flicked frequently to the man sitting at a ringside table. Carl Smiley: it was a nice name, and fitting, really, because he had a smiley face. Once or twice she met his eyes and he grinned and inclined his head slightly to show he was enjoying her performance. Around them the club throbbed with noise, music and conversation, the tinkle of bottle on glass, the occasional forced laughter of one of the girls, but Elaine became less and less aware of what was going on around her.

'Touch me, you only have to touch me . . .' she sang, and held out her arms in Carl's direction as though in supplication.

24

14 January 1985

The same old news of the strike was on the morning radio: how many pits working, how many men gone back, how many arrested on the picket-line, the utterly useless rhetoric of Ian MacGregor and Arthur Scargill. Jenny switched to the anodyne Radio 2 and began to open her mail.

An invitation to her cousin's wedding at Easter . . . she must go if she could. Family occasions were rare and not to be missed unless it was unavoidable. Two bills, and an offer of a time-share . . . would she ever have leisure to waste on a time-share? She threw the brochure spinning towards the waste-paper basket and opened the final envelope.

In the old days she would have recognized Barbara's handwriting, but now she had to turn to the third page to check on the signature. She turned back to the beginning. 'I know it's been ages, but honestly there is so much social life here I don't have time to draw breath.' Same old Barbara, Jenny thought and read on:

However, out of sight is not out of mind. I think about all of you often and hope you are all OK. I heard your name on the radio the other day and it gave me quite a thrill to think I was your best friend. Am your best friend, I hope? I've had two cards from Euan. He seems as scatty as ever but he can afford to be like that. I wrote to Alan after his mother died and got a nice little letter back. Very formal, but correct for that situation. I suppose Elaine is still living hand to mouth in a garret somewhere. I only hope her poor little girl is not affected by it all. Guy says the regulations on single parents should be tightened up. Still, it's gone on since the flood and I suppose it always will. Kath must be suffering, too, but, as Guy says, if they let Scargill lead them by the nose they only have themselves to blame. The one great success of our group is Keir, isn't it? He's never off the box. Guy says he must be making a fortune. We girls were slow not snapping him up in the old days. You're still free, so get cracking.

Now for my news. The wedding is set for July. All the trimmings, grey tail coats, everything. You know I'd love to have you for bridesmaid, Jen, but Guy has three girl cousins so it has to be family only. They all went to public school so it's very pukka-pukka . . .

'I wouldn't have fitted in, thank God,' Jenny said aloud, skimming down the rest of the page. She would have to buy a present but, God willing, she'd be too busy to go.

Barbara and she had been inseparable when they were children, sharing toys, clothes, even secrets. She had told Barbara the first time she fell in love and envied her because she'd been the first to have a period. Now she thought Barbara ridiculous with her pretensions and her snide remarks. Wait till she heard that 'poor Elaine' was living in the lap of luxury in Euan's Hampstead pad. 'I'll write tonight,' Jenny vowed as the radio pips reminded her she'd have to get a move on.

She brooded about Barbara on the tube, but once in the office, with the guests arriving and half of them demanding that their intros be rewritten, she soon forgot her. What egos people had, she thought, as she tried to explain why intros must be brief and couldn't include every tiny achievement. It was a relief when she had them all seated and the first turgid exchanges, which she would cut later, were over. It was then that discussion really got going.

In the control room Jenny looked at the sound engineer and saw his intent face, as the argument about the strike raged back and forth across the studio table. It was going well. The four speakers were all impassioned, but none of them was strident, and they were evenly matched.

'Well done, Jenny,' the engineer said when she had intervened to tell the speakers she had enough on tape.

261

The presenter thanked them by name and then gave the listeners his closing spiel.

'That's fine, Chris. Very good. One of your best,' she said on talk-back and then went through to thank the guests.

They all filed out of the studio, euphoric now that it was over. In the hospitality suite Jenny opened bottles and filled glasses – gin and tonic, orange and Perrier, lager – and directed people to the tweed-covered seats that lined the walls. The room had the neutral tones of a dentist's waiting-room; even the pictures on the walls were bland.

'Water and whisky, please. In that order,' Paul Kovalski said. 'I've got to go on to a meeting.'

Jenny handed him his drink, one ear tuned to the other conversations in the room. It was her duty to make sure everyone was happy, but this man was a newcomer, a novice, and needed a little extra care. She raised her glass to him.

'Here's to many more successful broadcasts,' she said.

'Thank you.' He had nice eyes and when he smiled the rather ordinary face was much improved. He took a drink and then leaned forward. 'Was I OK? Honestly?'

'Yes,' Jenny said. 'Better than OK. I thought you were good by any standard. Considering it was your first net-worked effort, I thought you were brilliant.'

'You'd better be telling me the truth,' he said. 'My wife will listen when you put the programme out, and she pulls no punches.'

Jenny asked polite questions about his wife and children, wondering why she was vaguely disappointed to hear he was married. She never got involved with married men. Life was complicated enough without triangles.

'What's next week's topic?' he asked politely.

'Surrogate motherhood. Though that's provisional – we stay flexible almost up to going into the studio. Hence your late invitation.'

'I thought I must be a second or third choice.'

'No, you were first choice. You were recommended by your Whip. He was in last week and I asked about likely newcomers.'

'I see. Well, I'm glad I came. Any more good news?'

'No, I think you've had enough for one night. I don't want to inflate your ego and spoil things.'

Paul Kovalski smiled. 'Tell me more about the surrogacy issue, then.'

While she talked of the Kim Cotton case – the baby born to order a few days before to a British mother and a wealthy American whose wife could not give birth – Jenny felt strangely peaceful. There was an air of content about this man, a feeling of sincerity that was very soothing. The Whip had told her Paul Kovalski was one to watch: 'His father came to Britain in the war. Married a Welsh girl. Odd combination, but they've spawned a nice lad. He'll go far in the party.'

The name had stuck in Jenny's mind when the other names had escaped her; she had remembered Kovalski and invited him. She was glad she had.

Steven had scoured the woods around Belgate, but they were picked clean of logs – or twigs even. He moved out of the woods down toward the sea, making for the blast, the beach beneath the colliery where waste had been tumbled into the sea for more than a century. It came into view, grey and featureless, a testament to the power of coal to affect men's lives for good or ill. Nothing lived here, on this grey shore line. Steven stumbled on the tracks of the mineral line and cursed under his breath: damn this for a lark – a collier scratting for coal like a bloody waster.

He walked between the banks of shale, seeing a fossilized tree root here and there, once a plastic bag impaled on a thorn, quivering in the wind like a flag of surrender. Twisted wire and cable lay everywhere, and the sand was a fine grey silt rolling into a slurried sea.

He looked up, seeing the pit sentinel-like on the cliffs above. He had worked there all his adult life; now he felt a stranger to the place.

'Is that you, Steve?' He turned to see a man he had worked with as a marrer years before.

'Aye, it's me, Stan. This is a nice business, isn't it?'

'The buggers should all be shot, Steve. Not that they haven't got a case, mind – I'll give them that – but the way they've handled it . . . By Christ, they've played right into Maggie's hands.'

They moved forward, kicking at stones here and there, seeing no gleam of coal, talking of MacGregor and Scargill and the other figureheads.

'Your lad's doing all right,' Stan said at last.

Steven strove hard to sound nonchalant. 'He's doing canny. His mother complains she never sees him, though, that's the only snag.'

'Does he ever come up here.'

'Oh yes,' Steven said, hoping he would not be asked to cite dates. It must be eighteen months since they'd seen Keir – easily eighteen months. 'Yes, he's back and forward all the time.'

'He was a party member, wasn't he?'

'Still is. His subs is paid right up to date, you can ask the secretary. I always thought he'd go in for politics one day – that's what he studied for and it's all still there. But a lad has to make a name for himself.'

'One time when he's up d'you think he'd come and do us a turn at the club? You know what lasses are like for a TV personality, and we're having a struggle making ends meet while this lot's on. If your Keir was to turn up we could advertise it, draw folks in. And if he's a good socialist he should support his Labour Club, shouldn't he?'

'He is and he will, Stan. You can depend on that.'

As they trudged along the beach Steven filed the worry of asking Keir to visit the Labour Club at the back of his mind: he would worry about it nearer the time. He went for a half of beer but left the pub at half-one. Cissie had been due back from the soup kitchen at one, so with a bit of luck there'd be something on the hob by now.

He was cutting across the waste ground behind the pub when he saw the man, attracted by his wary efforts to escape notice. He recognized him at once: Billy Wilder, one of the first to go back to work at the pit and a marked man since. He was trying to decide whether or not to acknowledge a scab when he saw the other men gathered by the gap in the fence that gave on to the road. There were three of them, shoulders hunched against the wind, caps pulled down over their ears. He had seen them earlier in the Half

Moon, trying to spin out a half of mild, making bitter jokes about their lot. Raylor and Cumpson were troublemakers, but the Harkness boy was a canny lad in normal times. They had been brewing up for trouble in the pub; now it looked as though they were going to get it.

Steven pondered his own course of action. If they pounced on Billy Wilder they would still be roughing him up by the time he, Steven, got to them. What did he do then? Pretend not to notice or put the boot in himself?

He tried to remember what had been said about Wilder at the time – something about one of his bairns having a disease and having to be thumped on the back all the time to stop her choking. Cystic something . . . Cissie knew the name. He didn't hold with scabbing on any account. On the other hand, he could imagine what it was like to have a sick bairn and nee pay. Also he didn't like odds of three to one.

The three men were moving now, half-running towards the shrinking figure of the scab. He was in for a pasting and no mistake. Steven hesitated. It was none of his business and if he arrived home looking as if he'd been half-killed Cissie would surely finish the job. All the same . . . he was turning to go back the way he'd come when the three pounced. He could hear the sounds of knuckle on flesh and Wilder's cries of pain. He had heard worse in the pit many a time, but what he had not heard before was the baying of men in pursuit of quarry. It was a new sound and he didn't like it.

He hurried across the intervening space, seized the first man by the back of the jacket.

'Lay off it. That's enough! Think of the union, you bloody fool. This'll look nice in the paper.'

'He's a scab,' Raylor said, his face contorted. 'A fucking scab!'

Steven had pushed his way in between them now and spread his arms. 'A scab he may be, Joe, and I don't hold with that any more than you do. But there's a procedure for dealing with that, and you're not it. So hadaway out of it now before I loss me rag, an' all.' They hesitated. 'Go on, I said loss yersel's. You've half-killed the poor bugger. Don't start on me or you'll get more than you bargain for.'

His words were bold but his heart was labouring uncomfortably inside his rib cage. If they called his bluff, he'd had it.

They turned away at last and Steven tried not to betray his relief. He turned to look at Wilder, who was trying to dust himself down, a fresh graze on his right cheekbone. Other half-healed bruises blotched his face and Steven felt his gorge rise.

'By God, lad,' he said. 'I don't know what drove you to scab, but thou's paid a terrible price for it and no mistake.'

Kath forced herself to dust and polish, to attend to the ailing fire and see to the children. Whatever happened she mustn't break down in front of Dave, and he would be in from picketing any time now.

Day after day the battle went on outside the pit. The armoured buses were waved through by police, their drivers' faces hidden by ski-masks. Then came the few miners brave or foolhardy enough to go in on foot, who had to run the gauntlet of brothers and sons and former friends, spitting hate now, raining blows if they could, shouting, always shouting: 'Scab' and 'Bastard' and 'Fucking traitor'.

'It's hell on earth there, Kath,' Dave had told her more than once. 'If we didn't need the picketing money I wouldn't go.' She knew he had doubts about the strike, had had them all along. 'There should've been a ballot, Kath, that's always been the miners' right. Scargill robbed them of that. If Maggie Thatcher had done it there'd've been an outcry. As it is, he's supposed to be a ruddy saviour. Beats me!'

Musing, Kath had stood at the sink, gazing out on to the yard. Now, suddenly, she remembered her terror this morning when she had gone to the calendar and marked out the days. She *couldn't* be pregnant – not now, not in the middle of all this. And if she was, how would she tell her mother?

She looked round the little kitchen, seeing the frayed lino, the chipped draining-board, the place in the corner where the plaster had crumbled. They had been going to fix all those things: 1985 was going to be their year! Do the house up and sell it, and move on to a house with a garden and a tree and a place for the children to play: that had been the dream. The strike had dashed their hopes; a third baby would bury them forever.

She had to talk to someone. If only she could run to her mother, feel the strong, plump arms close around her. But even if her mother had still loved her, she had more than enough burden of her own with all the men in the house laid idle. And she would be sure to blame it all on Dave. In the end it would sound like he'd raped her, never mind drawn her into a strike.

If only she still had the telephone she could ring Jenny. Jenny understood. Lately they had made arrangements by letter for Kath to be in the box down by the Half Moon at a pre-arranged time so that Jenny could ring her, but the next date was not until Friday. She would have to hang on until then.

Hearing Dave's step in the yard, Kath schooled her face into a smile. 'Hallo, love. Been bad, has it?'

He was shrugging out of his jacket, holding it at arm's length. 'That'll need washing. I got too near the front and lads at the back were spitting. There's a gobbet or two on there . . . or it feels like it.'

She went to him, folding him in her arms, feeling the unfamiliar tension in him. He had always been so easy, that's what she had loved about him – his lack of tenseness about life and love and everything in general. But he couldn't provide for his wife and bairns now, and that had brought him low.

'Oh Dave, I love you, I love you.'

'Oi, hold on. The bairns'll hear you.'

'They're next door with Marge's bairns. There's no one to hear. Say you love me.'

'I can't say things like that – well, not standing up here in the kitchen I can't. Wait till later on and I might.'

But the last few nights he had turned from her, unable to make love. Kath started to laugh, thinking of the irony of his impotence coming now, when it was too late. Her laughter was turning to tears and then to laughter again and she couldn't stop it.

'Hey up, Kath! Hey up, what's the matter? Don't cry, bonny lass, don't cry. I *do* love you. You know I love you . . . you and the bairns are the world to me. We'll manage. It'll be over soon, Kath, and we'll start again. We'll both work, I won't stop you. It'll be all right, you'll see.'

But while she dried her eyes and smiled to comfort him Kath wondered if it would ever be all right again, and try as she might she couldn't shut out the picture of her mother's face, her lips forming those four loathsome words: 'I told you so.'

She got on with her work, smiling at Dave until he went to get a wash, even singing sometimes to show him all was well. When Cissie Lockyer arrived with a bag of potatoes and carrots she thanked her warmly and offered her a cup of tea.

'No thanks, pet, Steven thinks I'm down the canteen. I'd best get back and get something on for his dinner.' She had hesitated then. 'Are you sure you're all right, Kath, love? You look a bit peaky today.'

'I'm OK,' Kath said. 'I couldn't sleep last night. It's the cold I think. We're not used to doing without fires in this part of the world, are we?'

'You can say that again,' Cissie said, and hurried off. Watching her, Kath wished she had taken the older woman into her confidence, clutched her and told her everything. Older women knew things about falling wrong: there'd been no terminations in their day, so they must have had something. The phrase 'back-street abortionist' popped up in her mind, making her wince. Please God don't let it be true. She couldn't take much more.

'I think I'll take the bairns for a walk,' Dave said, when he came back into the kitchen at the sound of his children's returning voices. 'I'll go light if I sit around here much more, Kath. Put the bairn in the pushchair and I'll get our Brian's coat.'

She waved them off and then walked to the sink, resting her hands on the cold rim, dropping her chin to her chest, letting the tears flow.

'Kath?' It was her father, stooping through the door, his face concerned. 'What is it, pet?'

'Oh nothing, dad, and everything. The strike and the

strike and the strike.' A spasm of pain . . . or was it sympathy? . . . crossed her father's face. Two jerseys showed under his jacket and he wore a muffler but his hands were blue with cold.

'It's a bugger,' he said, 'and no mistake,' rubbing his hands together gingerly, as though they were sore.

Kath moved the kettle on to the weak fire and sat down, motioning her father to Dave's chair. 'What brings you here? I've been married nearly five years and this is about the second time you've come round.' She sat up suddenly. 'There's nothing wrong is there . . . with me mam or the boys?'

'No, nothing wrong except we're all in the same bloody pickle. No, I came round for a reason. Before I get on to it, I wouldn't let on to your mam . . .'

Kath was stung. 'Are you even forbidden to visit me now?'

'No, lass, nothing like that. I just don't want your mam getting upset. Well, more upset.' He was fishing in his pocket. 'Recognize this?' The coin gleamed between his finger and thumb.

'It's your half-sovereign . . . the one I had in me shoe the day I got married.'

'That's right. It was my mother's, your grandmother's. She got it when she left service, and she kept it for her daughter. Well, our Nora died young, so it came to me, and I always meant it for you and your daughter. But I've been thinking – it's daft to keep it lying around while bairns is hungry. So I want you to have it now, Kath. What you do with it's up to you. They fetch a canny bit, so they say.'

She came out of her chair, moving into his arms, uncertain arms at first and then comforting.

'There, there now. You tell your dad all about it and then we'll have a brew. This'll all come out in the wash, you'll see.'

Keir could see Annie on the far side of the studio, deep in conversation with the floor manager. She was always plotting against him now, the cow. He felt his face grow taut and made a conscious effort to relax. The audience were there, faces expectant: it wouldn't do to let them see he was riled.

He felt his hairline grow suddenly wet and beckoned to the make-up artist, standing with her kit just out of camera range. She powdered him and then flicked her comb to restore his hairline.

'All right?' he said, smiling to show it was not her fault. She had a nice smile, which reminded him of someone. Elaine? Yes, little Elaine with the black stockings and paper-thin winter coat. She was shacked up with Euan now, according to Alan. He must look them up some time. It didn't do to alienate old friends – they were rich territory for muck-raking journalists.

Annie was moving into her warm-up routine now. He would get a proper warm-up man as soon as he'd sorted things out. That chap out of *Hi-de-hi* was good, or some up-and-coming young comedian. It might be worth trawling some of the clubs for a likely lad. That was what a star needed, fresh talent. He couldn't carry it all on his own shoulders.

'Ready, Keir?' The floor manager was there, speaking to him, then bending to her mike to speak to gallery. Keir felt the familiar wave of pleasure as applause engulfed him, as he walked out into the spotlight, paused, looked round in feigned astonishment that anyone should be applauding him, and then raised his hands to stem the same clapping that the floor manager, out of camera range, was doing her best to whip up. It was late morning but they must make it appear to be evening, when everyone was relaxed and ready to enjoy.

'Tonight, ladies and gentlemen, we have a fascinating topic for you. Think if you will of the moment of birth . . . a baby emerging from the womb. A miracle. But who does that baby belong to? The woman who bore it, or the woman who commissioned and paid for it?' Surrogacy was the topic of the day, but no one would deal with it as well as he would.

He could see them responding to the spiel, see the experts in the front row looking smug. He would have to watch that doctor from Queen Charlotte's . . . too articulate by half.

'But it won't all be the cut and thrust of debate. We have music too, the Thunderbirds with their new hit single . . . and someone I especially want you to meet. But more of that later. You know me, I like to keep some of the cards close to my chest.' He smiled, forcing the dimple into his right cheek, feeling the muscle harden as he looked into camera and gave a slow wink, the intimate gesture he had decided would unite him with his audience. The fading sports star lined up for the final spot would come as an anti-climax unless he got in some really thrusting questions. In their eagerness to stay in the spotlight once their athletic prowess went, they would do or say anything.

The floor manager was fanning the applause but it would have come anyway. He had worked for it . . . all those hours of studying the tapes with Annie, seeing what worked and what didn't, choosing the colours that did things for him, having his teeth capped which had been bloody agony, having his hair teased through the rubber skullcap until he had wanted to scream out, so that highlights could be applied. It had been worth it, all of it.

'Now you know me, I'm a northerner . . .' They always rose to that line. He had a sudden vision of his mother sitting at home: she would know he hadn't forgotten his roots. He smiled as the autocue rolled up, feeding him the lines that he had learned to deliver as though they had just floated up into his mind.

Euan had long since introduced Jenny to the delights of Dim Sum. Now they sat in the Rupert Street restaurant sampling the steaming bundles that looked strange but nevertheless tasted wonderful.

'I don't mind telling you I'm worried, Jenny. Elaine's always been thin but now . . . Sometimes she picks L.J. up – and much as I adore her, the kid's a handful – and she sways with the weight in her arms. Sways!'

'The last five years have been tough for her,' Jenny agreed.

'And the Club de Naff is no picnic,' Euan said dolefully. 'We're classed as musicians, which means they can treat us a little less courteously than squashed cabbage-leaves.'

'What is the proper name of that place?' Jenny asked, grinning. 'When Bart and I came to see you that time I actually said 'Club de Naff' to the taxi-driver.'

'It's the Blue Fountain,' Euan said, 'but Red Light would be more appropriate. I have more than once had to interpose my body between one of the ladies and an over-importunate client. They think the girls are up for grabs, which of course they are. Which brings me back to Elaine . . . I'm relying on you to think of some way to fix it, Jen. I can supply cash unlimited – well, relatively unlimited. I just can't work out the logistics of how to free Elaine for a while.'

'You mean for a holiday?'

'Yes, in the first place. My plan is that Elaine should have a break for a couple of weeks – somewhere in the sunshine, if that's what she wants. When she comes back, we don't reunite her with L.J. at once, we go into rehearsal with Carl, doing the songs I've written for her. Once we've got the set right, I think we can move on from the Club de Naff. Or up, to be more precise. So, two problems: where do we send L.J. to be happy and cared-for; and where can Elaine go for the same treatment?'

Jenny smiled. 'I can guess where Elaine would like to go. Remember when Barbara and I went to Jersey? Elaine has always hankered after it. "Tell me about Jersey," she used to say, "I'll go there one day." We'd have to ask her, but I bet she'd plump for Jersey – as long as she needn't look up Barbara.'

'Perfect,' Euan said. 'Not an exhausting journey, so she can get the maximum effect from her stay. And Barbara doesn't need to know she's there. What about l'enfant terrible?'

'I think I've got an answer to that too,' Jenny said slowly. 'Would you be able to pay for L.J.'s keep . . . well, I'd chip in, anyway?'

'Of course. No problem.'

'Leave it with me. I'll get back to you. It just might be a blessing in disguise if we can get it all to work. Will you be going on holiday with Elaine? She might be lonely on her own?'

'I'll go if I'm needed,' said Euan, 'but I have a shrewd

suspicion that someone else may be ahead of me. Carl, the piano-player. He seems smitten.'

'Is he good enough for her?'

'He's a good bloke, Jenny. I think he'd do very well, and he likes Her Majesty, which is an essential requirement as far as Elaine is concerned.'

Jenny wiped her mouth, seeking the right words. 'I used to have fond hopes of you and Elaine getting together?'

Euan smiled. 'Ever the romantic, Jenny. No, there *was* someone once but it's over now, and I don't think it will ever happen to me again. But I'm very fond of Elaine, partly because of who and what she is, and partly because L.J. is Jed's child. I'll always stand by them.'

Who had Euan loved? Someone in Zimbabwe? Before that there had only been Kath or Barbara, and before them he had been a kid. Jenny was plucking up the courage to probe a little when a shadow fell across the table.

'What a coincidence, bumping into you two! I could hardly believe my eyes.'

'Alan!' Euan was standing up to embrace him. Jenny took in the expensive suit, the same still features, half-smiling now as he bent to kiss her cheek and then turned to introduce the tall blonde at his side.

'Jenny Sissons . . . Eleanor Cardus. Jenny and I are old friends, Eleanor. How are you, Jenny? Euan tells me of your broadcasting triumphs occasionally, but we ought to get together more.' There was a faint flush of embarrassment on his cheek and a note of pride in his voice when he had introduced Eleanor. So Alan was in love, Jenny thought, and was glad. He was meeting her eye now and she knew he was remembering the tiny sitting-room with the trappings of a funeral around them.

Eleanor Cardus laid a long, slim and proprietary hand on Alan's arm. 'Yes, you *should* get together more often. Old friends are precious. Do something about it, darling, or I shall nag.' She was smiling at Jenny, the wide, benevolent smile of the lady of the manor doling out jelly to the starving tenantry. 'But now, please excuse us. We have a chambers meeting in an hour. It's such a pity . . . I've wanted to meet you all so much, but we really shouldn't have taken time out to eat, not today. Chambers meetings are not something we can treat lightly.'

'Of course,' Jenny thought, seeing the black skirt, the white silk blouse under the pale leather coat, 'she's a barrister, too.'

'Well, well,' she said, when the two had moved on to the door. 'That was a surprise.'

'I've met her before,' Euan said a little ruefully. 'I didn't mention her because I hoped she was a fleeting *mésalliance*. I may have to reappraise the situation, though.'

'Why?' Jenny said.

'I need notice of that question. I'm still trying to adjust to hearing Alan called "darling" and watching him blush. I mean, he's one of the best, but romantic he's not. "Darling" – hell's flaming teeth!'

'I don't know,' Jenny said thoughtfully. 'The thing about Alan is that however close you come to him you never know what's going on in his mind. He's not secretive . . . not exactly. He's . . .'

'Deep,' Euan said lugubriously. 'Deep. Now, if I can catch the eye of that nice little girl with the hair slide, we can have some vine-leaf bundles.'

Jenny kissed him warmly when they parted and then hurried back to her office to tie up loose ends. She wanted to get home and write to Kath, and the sooner the better.

When she was finished she came down in the lift and walked across the marble entrance of Broadcasting House. She was almost at the door when she heard her name called.

'Jenny . . . it is Jenny, isn't it?'

She turned to see Paul Kovalski, the young Labour MP who had taken part in last week's programme.

'Hallo. Yes, it's Jenny. I'm surprised you remembered.'

He grinned. 'Remembering names is marginally more effective than kissing babies.'

'And not as messy,' Jenny suggested.

'True,' he said, following her on to the pavement. 'Will we get a cab?' But Jenny was already signalling furiously.

'Where are you going?' she asked as it drew up.

He shrugged. 'Nowhere special. Somewhere to work off the surplus adrenalin. I've been doing something for the six o'clock news.' He paused. 'You wouldn't like a drink, would you?'

Jenny had wanted to hurry home and write to Kath, but half an hour wouldn't hurt and she knew what it was like to come out of a radio studio with adrenalin running in your veins like firewater.

'One drink,' she said. 'You pick the pub.' But in the end it was she who gave a name to the cabbie: 'The Duke of Wellington in Portland Street.'

'I really don't know London yet,' Paul said. 'Not from the social point of view. My wife and family are still back in the constituency, so I don't go out much.'

They settled in a quiet booth, almost the only customers at that time in the afternoon. In an hour or so it would fill with home-going workers, but now it was peaceful.

'Tell me about your piece. Who was it for?' Jenny demanded, suddenly glad that she had dressed up for her lunch date.

'I was talking about the trouble the pound is in,' Paul answered. 'I never quite caught the producer's name . . . he wasn't as charming as you. The Government's taken emergency action to halt the slide, but it hasn't worked. It closed tonight at a new low of 1.1105 dollars.'

'That's bad?' Jenny asked, smiling at the compliment.

'Very bad. My problem tonight was that I really didn't want to talk it down further, but politics being what they are . . . adversarial . . . I've probably pushed it down another point.'

'If you hadn't, someone else would've done,' Jenny said comfortingly. 'You were probably more constructive than most.'

'I'm glad you said yes to a drink,' Paul said. When he smiled his eyes crinkled, making him look like a boy.

'Do you miss your children?' she asked, and waxed suitably enthusiastic over pictures of a solemn boy of seven and a smiling girl of five.

'How did your programme go . . . the one on surrogacy?' Paul asked when the pictures were put away.

'I was pleased with it. Violent opinions, of course, but also some attempt to be dispassionate and see the way

275

ahead.' She leaned forward. 'The Cotton baby's gone to America now, did you know?'

'Yes,' he said. 'Do we pity it or envy it?'

'Well,' she said, shrugging out of her coat and leaning forward on the table, 'now that you've asked me . . .'

25

29 January 1985

Keir toyed with his smoked salmon and kept his eyes down.

'You won't get me to say a word against Annie,' he said. 'I owe her everything. Whatever she is . . . well, anyway, I don't want to make things worse.' He saw the barb had gone home and changed the subject.

'It's been a bad week for broadcasting. James Cameron dead, and Lord Harlech.'

Across the table the boss of Global Television nodded. 'Both good men in their time.'

'Absolutely. Cameron was my idol when I was starting out. Such authority.'

Keir broke open a bread roll and furrowed his brow, trying to look loyal but troubled. He must let it be dragged out of him: it wouldn't do to say, 'Annie goes or I go.' He was fairly certain he would win such a contest, but it would be remembered. 'Softly softly catchee monkee' was a much better policy.

'Have you and Annie talked about the new series?'

'Yes,' Keir said cautiously, 'we've thrown a few ideas around. I'm afraid . . .' He let his words tail off.

'Whose ideas?'

'Well, mine mostly. Not that Annie doesn't have marvellous ideas . . . well, you know that, you've worked with her for *years*.' Just the right amount of stress on the final word.

'Yes, Annie's a byword in the business.' The mogul pursed his lips. 'A character . . .'

Keir frowned slightly in what he hoped was 'young man led grievously astray' style. 'Yes,' he said. 'She's been . . . well, I owe her a lot . . . which is why . . .'

Jesus Christ, this was going to take forever. *'Move the fucking slag to "Children's Hour" but get her off my back,'* was what he wanted to say. And if something didn't happen soon he just might!

Their veal was placed before them and the Niersteiner Gutes Domtal Keir had asked for specially.

'Ah . . . good,' he said, sniffing at it appreciatively. Then he leaned forward and spoke confidingly: 'It's all a bit hard to take in, you know. I mean, I come from mining stock – salt of the earth, but we didn't go in for much wine. And here I am drinking this, with you . . .' He smiled again. 'And I've got the best job in the world, and the ideas are flowing like the Flood . . . it's bloody marvellous, I can tell you.'

It was mention of ideas that did it. He saw the opaque yellow eyes flicker at the word, and then there was a long moment of mastication before the older man laid down his knife and fork.

'If I were to take Annie from you, for something I have in mind . . . could you give rein to these ideas of yours?'

He'd cracked it! Keir thought of the moment when Annie was told she was out – shunted sideways if not down. God, he'd love to be the one to tell her she'd been shafted!

'Well,' he said carefully, 'it would be marvellous if I could just expand a little . . .'

Jenny had shunned the canteen at lunch time. She'd had enough of the miners' strike, which was all anyone seemed to talk about at the moment. There was something almost obscene about the way people pontificated about it, sporting their 'I'm Supporting the Miners' badges, without taking the trouble to discover and understand the root cause of the misery.

It was too cold for the park so she sought out a side-street sandwich bar and sat thinking about L.J. and Elaine, who would be speeding north now in the piano-player's car. Jenny had written to Kath as soon as Euan had spoken to her, suggesting that Kath should take L.J. for a few weeks. *'I know it's asking a lot, Kath, but Elaine needs the break. She can afford to pay (Euan's shout) so it might help you a bit, but the important thing is that Elaine knows L.J. is with someone she can trust. Otherwise she won't get any benefit from the exercise.'*

There had followed a spirited conversation on the phone, by arrangement, when Kath had said she couldn't take money and Jenny had abandoned the restraint of a lifetime and sworn fluently down the receiver.

In the end it had been fixed. Euan would pay £25 a week, ('half of what he'd pay elsewhere,' Jenny had pointed out). L.J. would stay with Kath and Dave and their children for three weeks, and Elaine and Carl would go to Jersey for two weeks and spend the third week rehearsing Euan's new set.

'Wunderbar!' Euan had said when she told him it was fixed. 'You're a genius, Jen.'

Now she sat in the sandwich bar imagining Kath at the door, welcoming them, drawing them inside. If only she could be there, too, to see it.

'Hallo. It's Jenny Sissons, isn't it?' Jenny remembered the girl at once: Toni, the researcher who had brought Keir to London and been abandoned for her pains. 'Can I join you or do you want some peace? Just say if you do. I know what it's like,' Toni said tentatively.

'No, please. I'll be glad of the company.' Jenny slid along the leatherette seat and put her bag on the floor.

'How long is it since we met?' the girl said, considering.

'It's five years ago', Jenny said, 'that you came to New-castle. It was my last year at university, so I can date it exactly.'

'Keir's done well, hasn't he?' Toni said, lifting the top off her roll and grimacing at the contents. 'You know we lived together?'

'Yes,' Jenny said. 'Yes, I did know.'

'He's a bastard,' Toni said, biting into her ham and salad. 'I didn't have a marvellous opinion of men when I met him, but he took even me by surprise. He is an utter sod. He's with Annie Love now, but you watch – he'll shaft her sooner or later.'

'His ratings are good,' Jenny said, wanting to be at least a little loyal.

'Yes,' Toni said. 'He's got the gullible-granny section,

the matinée-idol crowd. Have you seen him pose? I watch him scanning the audience – you watch and you'll see it too – and he picks out someone to make a good picture. No one brilliant, because that would detract from him. He marks them out, and then he'll go down on one knee and turn just a fraction to camera. God, it makes the flesh creep. He's had five researchers in two series.'

'Why?' Jenny said.

'Because he disregards every blind thing they tell him. He's not interested in the show, the whole concept . . . it's just a backdrop as far as he's concerned. No one worth their salt'll work in that atmosphere for long.'

'I don't think you like him much,' Jenny grinned.

'The only way I'd like Keir Lockyer is hung up by the balls over a hot fire. I'd like that all right. Still, I'll see my day. You know the old adage: "You can fool some of the people . . ." Just wait! I'm going to have coffee – can I get you another?'

Kath had talked it over with Dave again and again before they finally agreed to take Elaine's child. At first he had been against it: 'We've got enough on our plates, Kath. We can hardly feed our own at the moment, never mind someone else's.' She had wanted to point out that £25 a week would feed all three children, but she didn't dare. If he could feel he was doing a favour he might agree. If he felt he was being done one, he'd refuse it.

She won him over in the end by pointing out that Jed would have wanted L.J. to come to them. She felt mean about the lie, but it was in a good cause. 'Belgate's her roots, Dave, and you liked Jed. If he'd still been alive, none of this would have happened. Now Elaine's got to have a break . . . Jenny says it's desperate . . . and it's only for three weeks.'

Euan's cheque had been for £100, 'just in case', and Kath had cashed it at the Paki shop because their joint bank account was overdrawn and under sanctions. It had enabled her to stock the pantry again, to put something by for bills, and, best of all, it had reprieved her father's half-sovereign. When L.J. went home it might have to go again, but for now it was safe in Kath's dressing-table drawer.

She was curious about L.J. 'A little madam,' Euan had called her in one of his letters, but he was obviously besotted with her so she couldn't be that bad. Jenny had called her 'basically loving but strong willed', which summed up just about every child Kath had ever met.

'OK, so she may be a handful,' she told Dave. 'We can cope for three weeks.' He had looked dubious and muttered something about the effect L.J. might have on Brian, but that was all. So far so good.

She was unprepared for her first sight of L.J. The child was dark-eyed and fragile, so unlike her own chubby Margaret. She had on a beautiful hand-smocked dress which almost came to her ankles and had obviously been bought by a man . . . Euan presumably . . . and there was a string of golf-ball-sized pearls around her neck.

'Hello,' Kath said gravely to L.J. And then held out her arms to Elaine. 'Come here, you. It's been far too long.' She hugged the thin body, still that of a child in spite of motherhood, and then held out her hand to the man bringing up the cases in the rear.

'I'm Kath.'

'Carl. I'm a friend of Elaine's.'

'I'm sorry, I should have introduced you . . .' Elaine said, and then Kath was sweeping them into her house, to offer them tea and cut into the first fresh cream sponge they had had in the house for nine months.

Brian eyed the newcomers warily at first but he was bigger than L.J. and on his own home ground. Before long they were seated on the floor playing with his Mousetrap game, L.J. issuing the orders, he gazing at her in admiration.

'She'll boss him around if she gets away with it,' Elaine said reprovingly, but her eyes said, 'Isn't she wonderful?' and Kath felt a wave of fellow-feeling. Tough as things were, she was here in her own home with her children and no need to farm them out, not even for a day. She watched her tongue, being careful not to mention Dave too often.

'He's down at the blast for coal,' she said when Elaine asked, 'but he'll be back before you go.'

They settled either side of the fire, Carl between them presiding over the tea-tray because they were too busy with catching up on one another's news to pour tea efficiently. They tried to include him in the conversation but he just smiled and relaxed, watching the leaping fire, seeing the shabby home for the comfortable, lived-in place it was. The children played boisterously in the corner, over their initial shyness now and relishing having company.

'I've told L.J. what's happening,' Elaine said, as the clock moved on towards the time they must leave.

'Is she OK about it?'

'I think so. I said a few days and I'd be back: we're only going to Jersey, so if it doesn't work out I can always come back and fetch her.'

'Jersey!' Kath said. 'I envy you.'

'I've dreamed about it ever since Jenny came back, I don't know why. Mebbe it's the names . . . Grève du Lecq, St Brelade's Bay, L'Etacq . . . and the flowers, the pink lilies. Jenny should be a writer. She made it all sound wonderful.'

'She might be going to do travel,' Kath said.

'I know. On TV. It'll be funny if she gets in a programme with Keir, won't it?'

'We can claim we know someone else famous, then,' Kath said, and, turning to Carl, 'Have you met Keir Lockyer? He comes from round here.'

'So Elaine told me. No, I know of him, of course – who doesn't by now? – but we've never met.' He looked at his watch. 'I don't want to hurry you, Elaine . . .'

'Carl's nice,' Kath whispered, in the flurry of leaving and saw Elaine's eyes light up. They were going to be happy: thank God, Kath thought. And then Dave was striding in at the yard door, sweaty and grimed but holding out his hand first to Elaine, then to Carl.

'I'm proud of him,' Kath thought. 'I really am proud of him.'

Elaine's eyes filled at parting and L.J.'s intense little face became grave.

'Tell you what,' Dave said, holding out arms he had wiped on the sides of his sweatshirt. 'You come to me and we'll wave your mam off together.' The child hesitated and then moved forward and Dave lifted her up. 'Poor little bugger,' his eyes said.

Kath kissed Elaine and bundled her into the car.

'We'll be back for you very soon,' Carl said to the child, and put the car swiftly into gear.

'Best way,' Kath thought and liked him even more. L.J. had suddenly drooped and a tear was glittering in the corner of her eye.

'Come on, then,' Dave said when the car was out of sight. 'I might take you and our Brian to see some pigs.'

L.J. looked at him. 'Real ones?'

'Yes.'

'Go on,' she said in tones of disbelief.

'You go on,' Dave said. 'If you don't believe me, I'll show you.' He helped her into her coat and accepted her proffered hand. It was going to be all right, Kath thought, wishing the phone still worked and she could tell Jenny all was well.

Keir heard Annie come into the flat and made a mental note to get her key from her. The watercolour above the fireplace was hers, and he'd have to give it back if she was naff enough to ask for it.

'Hi.' She was pushing those curious fingers of hers through her hair and flopping into the couch. God, that he had let those hands fondle him! He had become obsessed with them lately, hating the large knuckles, the veins on the backs, most of all the sinews. She was a leathery bitch.

'How was your day?' he asked sympathetically, handing her the statutory gin with a twist of lemon and tonic.

'Vile. I loathe having to negotiate when I'd really like to issue diktats. Still, I want to change the format a little for the next series . . . we'll talk about it later, but I've lined up Schaeffer to direct.'

'Oh, good.' He turned back to the bottles, wondering if her word would be binding once she was out. He wasn't having Schaeffer at any price.

'What've you done today?' she asked. He turned back to answer.

'Not a blind thing. Read my brief, talked to wardrobe, had a session on the sunbed . . . otherwise zilch.'

'Eat out?' she said casually. Did she know something?

'No, just sarnies in the canteen. Shall we go out tonight?'
'I think so. I couldn't face a stove. Where do you want to go?'

For a wild moment he thought of taking her back to L'Escargot where he'd lunched today. It would be exciting to fend off any remarks from the head waiter about seeing him twice in one day. But it wouldn't do: he mustn't rock the boat now, when it was almost accomplished. 'You choose. You always make the best choices.'

He waited while she changed and then held her jacket for her to slip into. 'Love you,' he murmured against the angular cheek and held her arm all the way down to the car.

L.J. declined a bath very politely, like royalty refusing autographs. 'Not tonight, thank you.' She consented to hang the large pearls on a drawer handle. 'They're my mammy's,' she told Brian. 'You can't touch.'

'They'll be there for your mam when she comes to get you,' Kath said. She had feared tears at bedtime, but so far so good.

'Can the man read?' L.J. said, when she was tucked in beside Brian.

'Yes,' Kath said, trying to hide her smile. If madam thought Dave would read bedtime stories for her, she was in for a shock. She had sat between his knees to watch television and had persuaded him to kneel on the floor to play Mousetrap, but that had been pushing things to the limit.

'Can he come?' L.J. asked as Kath opened *Paddington Bear*.

'Who?' Brian said, becoming frustrated with the conversation.

'Your dad,' Kath said. 'She wants him to read a story.'

Her son's face said it all, but Kath toiled obediently out to the landing. 'You're wanted,' she called down the stairs.

'What for?'

'Just come up.' A few moments later, after Dave had scorned *Paddington* and opted for the *Dandy* annual, Kath tiptoed downstairs and set about making his supper. Wait till she saw Jenny!

'Before you say anything,' Dave said, when he came downstairs, 'I only did it because she's a bairn in a strange house. I'm not making a habit of it. It's women's work, reading to bairns.'

Kath dried her hands on a tea-towel and laced them behind his neck. 'You're a bloody fraud, Dave Gates. She's got you eating out of her hand, don't deny it. I'm jealous.'

'No need,' he said, moving his hands down to her waist and lifting her slightly till their faces were level. 'Only one woman for me. It's a pity she's such a tartar.'

Kath was tempted then to tell him about her suspicions, but it wouldn't do to spoil the moment. Not tonight, when things had gone so well.

'We have a chance to miss L.J. when she goes,' she said later, as they sat with their plates on their knees. She had hoped he wouldn't notice they were having lamb chops, unheard-of luxury, and he was chomping away without asking how they had been paid for. 'Yes, we'll miss her,' she said again.

'We might,' he said. 'But two bairns is enough for me. I don't want any more, thank you.'

Suddenly Kath was reminded of the child growing within her, and the chops, so long desired, turned to ashes in her mouth.

Steven feigned sleep as soon as Cissie came into the bedroom. Once it had been sweet to watch her getting ready for bed, hear her downstairs in the bathroom, draw her soap-scented body into his arms, to rest his lips against her hair. But that could lead to things and lately . . . lately he had become afraid of the quality of his performance. He could still do it but there was anxiety now, ticking away at the back of his mind. He was sixty-two, after all.

Beside him he felt Cissie stir, curving herself against him. She wanted it, he could sense it. If only he could be sure of himself!

He thought of all the years he had taken his prowess for granted, thrown it away almost, and suddenly he was frightened. It was all closing in on him now: no pit, no son – or as good as! All the areas of his life that had mattered. He had sold his job, and now he was being punished for it. There would be no jobs for Belgate lads soon, and he was one of the sods who bore the blame.

He felt a sob grow in his chest and grew terrified that it might escape. He turned then, seeking cover, and took Cissie in his arms.

'All right, pet?'

'I suppose I must be. Comfy?'

'Aye, I'm comfy.' He put his lips to her temple but was careful not to move them to her mouth. 'Just a bit tired. We'd best get to sleep.' He kissed her again and then turned his back on her, letting his tears fall silently to his pillow.

26

1 February 1985

Jenny got a seat on the tube, a rare luxury. She sat for a few moments, still blinking sleep from her eyes, and then began a cautious perusal of her morning paper, being careful to keep her elbows in for fear of annoying the passengers on either side. She came across Alan's article near the back, a half-page setting out the legal position of miners on picket duty. From her cursory read-through it seemed a carefully argued piece, and she felt a glow of proprietary pleasure.

She glanced at her watch as she walked across Portland Place: time for five minutes' peace. All Souls was quiet at this time of day, with only two others reverently bowing their heads in far-apart pews. She slipped into the back row and closed her eyes, feeling the silence enter her. She was not particularly religious, had not been so since Sunday School days, but there was something about a church that got to you.

When she went back out into the bleak winter sunlight she felt refreshed. She went through the portals of Broadcasting House, flashing her pass to the commissionaire, determined to do her best that day. 'And nation shall speak peace unto nation,' the motto said: that was a big concept, but at least she could try to speak truth in her work and make it intelligible to all who heard it.

Perhaps she could do something else on the strike, if she could just think up a fresh angle? It had dragged on for so long now that it was difficult to find a peg on which to hang a programme that hadn't been used a dozen times before.

She was at her desk, toying with the run-through for her next recording, when the phone rang.

'Jenny?' It was Paul Kovalski.

'Paul. What can I do for you?' She felt suddenly irrationally pleased to hear his voice.

'Oh nothing . . . I just wanted to say I'd be going north this afternoon as usual. Perhaps we could get together when I come back on Monday?' There was doubt in his voice and Jenny knew he was ashamed of what he was doing – a married man making overtures to a single woman. But she knew the score. If something came of this it would be of her volition and she would have to accept the inevitable consequences.

Even as she made polite conversation she was marvelling at the easy way she had accepted his being a married man. Normally she would have shied away from any such involvement. Now she was not only acquiescing in it, she was encouraging him.

'We could go to a show, perhaps,' he was saying. 'I can't work up enthusiasm for going to the theatre alone.'

'That would be lovely, Paul. Have a good weekend and we'll talk about it on Monday.' Long after she had put down the phone she was wondering just why she had behaved as she had done.

'There now, here's a nice little girl come to meet you.' Steven looked up as Cissie ushered a child into the kitchen.

'She's a funny little thing,' he said, looking at the child's dungarees, covered in gaudy patches, and the string of big pearl beads around her neck under the fur-trimmed anorak.

'Sh, Steven, don't upset her.' She bent to remove the child's coat. 'This is Lavinia Jane. She's come to stop with Kath Gates for a bit while her mammy's away. This is Steven. Say hello to him, pet.' But the little girl was moving towards the fireplace, looking up to the big pot dogs on the mantel ends.

'Can I see?' She said it imperiously and then added, 'Please.'

'Well, get them down, Steven,' Cissie said. 'I'll just get her a biscuit.' She lowered her voice for Steven's ear alone. 'She's missing her mam. No dad. I've brought her out to give Kath a break. She's a funny little thing if you get her on talking.'

Steven lifted down one of the waaly dogs and sat down in his chair. The child came to him confidently enough, moving between his knees to get closer to the ornament.

'Funny,' she said, pointing to one brightly painted eye. It felt strange to have a small child in such close proximity after so many years. He thought of Keir at that age: he'd been a bonny little lad, bright as a button. And he too had liked to play with the waaly dogs.

He drew the ornament back and then suddenly advanced it. 'Woof, woof,' he said and grinned as she squealed with delight. He lifted her on to his knee, marvelling at how light she was, like feathers.

'How old are you?' she asked. Cheeky monkey!

'As old as me tongue and nearly as old as me teeth. How old are you?'

'I'm five in August and I'm called Lavinia Jane Gentry and I live at 72 Vale Street, Hampstead, and if I get lost the phone number is 917 4342.'

'That's a mouthful,' he said, amused.

'Yes,' she said, tiring of the dog and turning her attention to his braces. 'What's this?'

He pulled them out and let them snap back. 'Pop,' he said.

'Again!' she demanded, and he obliged. He looked up to see Cissie gazing at him and the child, a mug of orange in one hand and a biscuit in the other.

'If only our Keir had had a bairn like this,' she said, and held out the mug to small, eager hands.

Alan had to walk carefully up the garden path, which was covered with dead leaves, converted to slime by the rigours of winter. If he kept on the house he would have to arrange some kind of maintenance. One of the nets had come adrift in the bay window and all the curtains looked suddenly grimy and unkempt.

He turned the key in the door and stepped over the small mountain of mail on the mat inside. Circulars mostly; he had had all worthwhile communication redirected. He carried the envelopes through to the kitchen and dropped them one by one into the bin. The house felt deathly cold and yet strangely airless, as though a heatwave had taken the oxygen away.

The water pipes rattled when he turned on the tap but the gas ignited with a satisfying plop and once he saw the flame he felt suddenly warmer and more welcome. He made a cup of black coffee and carried it through to the bathroom. If he was going to stay here while he was at Durham for the Assizes, he would have to warm the place up.

He looked around the room, seeing the familiar objects suddenly ghostly and strange. Had this cold, comfortless house driven his mother to take her life? But in his heart he knew that there was much more to it than that, that the seeds of his mother's sadness had been sown much earlier, perhaps even before he was born. He picked up his father's photograph, trying to remember the time when they had all been together. Had they been happy? Yes, they had. He could remember fire-lit evenings, the peace between his parents, kisses dropped by one on the other's brow, hands held, hugs at partings. So what had been the problem?

He had once considered the possibility that he was a bastard, that one or the other of his parents had been unable to make a legal marriage, but the certificate had been there in the bureau along with all the other family documents, so that had not been it.

Alan felt a sudden terrible melancholy and not even the thought of his cases next week could cheer him. Was the Bar really what he wanted? Now that he did not have his mother's ambition to satisfy . . . he paused, thinking. It had not been just an ambition on her part, it had been an obsession, and now he was free of it. He could leave the Bar if he chose and try to build a new career as a journalist.

He knew he had made a good start in his first profession. Work was beginning to come in, for having a pupil-master with a thriving practice in crime and the common law had been a huge advantage. His place in busy chambers was evidence that he was well thought of, but did he have it in him to be more than a hack with a run-of-the mill practice? To be a top-flight barrister you needed a first-class brain, a mastery of the law, and an insatiable appetite for long hours of gruelling hard work. You needed detachment, too; even a streak of cruelty. Did he have those qualities? Sometimes, prosecuting, he became so sorry for his victim that it required all his powers not to start arguing the opposite case. And yet he could be hard, cruel even, when it was necessary.

Alan looked across to the picture of his mother, standing behind his father's chair. Had he repaid her years of effort and self-denial? Probably not. He got to his feet and went upstairs to hang out his black jacket and striped trousers, and put the black tin box with its gold lettering, containing his barrister's gown and wig, ready for Monday morning.

There was a small framed photograph of Eleanor in his case and he set it on the dressing-table. What would the cool Lady Eleanor make of this modest little house. 'Lady Eleanor' was her nickname in chambers, a tribute to her poise. Could he bring her here, to a house more suited to the '20s than the '80s? Jenny had understood . . . but then Jenny was Jenny.

He crossed to the window, seeing the tip of the hill behind the rooftops where the eight of them had stood five years before, tempting the fates with their bragging. How many of them would gather there at the end of the decade? Alan was struck by the sudden thought that if Keir turned up he would surely bring half the world's press with him. The ridiculous concept cheered him enormously and he was whistling as he went downstairs to find the makings of a fire.

Keir sat with his back to the wall, surveying the other tables. Two MPs sampling the fleshpots: well, they could afford it. Stuart Copeland of Police with another man and a striking blonde; two people whose faces he recognized but whose names escaped him. Everyone else here would be in business; they were mostly men, mostly guzzling.

He turned his attention back to the man opposite. 'I like this place. Ted Lundgren brought me here first, and I was *very* impressed!' He smiled disarmingly. 'I was young and

raw. I looked round and saw one or two famous names
. . . I couldn't eat! I was awe-struck.'

The man opposite nodded. 'I know what you mean.'

Keir leaned forward, reaching for the slim stem-vase and
seeing how good his nails looked now. He moved the vase
a fraction. 'Well, let's get down to business. I want to turn
the show around. OK, it's reasonably successful and the
ratings are good, but I want mega-good ratings. I want to
be on the first or second page of the national papers, not
only the tabloids, every week. Not for the usual trivia but
because we're hitting the target . . . week after week after
week. I've got my audience – no boasting, but it's there –
and now I want a *bigger* audience. I want seventy-five per
cent of the available audience. What can you do to help
deliver?'

He listened, nodding, as the guy went into his spiel. He
came recommended, but Keir had doubts. He was too
painstaking, too careful in his exposition. What was needed
here was a buccaneer, someone as ready as he was to take
a gamble.

'Well,' he said, finally, as he signed the credit card slip
proffered on a plate by a hovering waiter, 'let's both mull it
over. I'll be in touch. Now, as I told you, I've got this
frigging little quiz-show to do, so I must split.' He laid his
crumpled napkin on the table and stood up, giving the
man his famous smile.

As he walked between the tables towards the door he
was aware of one or two glances, a muttered, 'See who
that is.' If he could turn heads in here, he was moving up.
Onwards and upwards. He slipped the doorman a pound
coin for hailing a taxi, and gathered his coat around him to
climb inside.

'Thames TV, Euston Road,' he said and settled back to
enjoy the drive.

Euan was waiting at the kerb when Jenny emerged from
the huge gold doors of Broadcasting House, the motor of
his XR3 idling.

'Thanks, Euan. This is bliss.' Euan leaned across her to

secure her door, and then threaded his way expertly into the five-thirty traffic. 'Is Elaine excited?'

'Like a dog with two tails. She has a few tears every now and then when she thinks of the Princess, but Kath rang at lunch-time, and that really bucked her up.'

'Good for Kath. Did she say how she was managing?'

'No, I think she just said L.J. was fine and Elaine shouldn't worry. I had a word with her before her money ran out, asked was she OK for cash, and she said yes. But she says that anyway.'

'Hopefully, the strike will be over soon,' Jenny said.

'Not if Scargill has his way,' Euan said, grimly.

Elaine was surrounded by clothes when Jenny entered her bedroom.

'Thank God you're here, Jenny. Look, you tell me what to take. Will it be hot or cold? Shall I take a mac? Just pick for me, I can't make up my own mind.'

She sat on the window seat while Jenny made an inroad into the pile, folding a garment here and there and packing it.

'I still can't believe it, you know,' Elaine said. Her thin wrists jutted from her mohair sweater, and her hair was looped up in a pony-tail with one of L.J.'s bobbles to secure it. She looked about sixteen until you saw her tired eyes.

'Well, it's true,' Jenny said firmly. 'You are off in five minutes and I hope you're not going to leave this mess behind when you go.'

Elaine obediently started to clear up the confusion, turning suddenly and reaching out to hug Jenny. 'You've been a real brick to me, Jen, you and Kath and the boys. I've got Jed to thank for getting you for friends and I don't forget it.'

'Is this a private love-in or can anyone participate?' Euan asked from the doorway, holding a tray of coffee and pizzas in his hands.

Elaine took the tray from him and put it on the dressing-table. 'Come here and get a proper hug.'

Jenny watched them, wishing once again that they might wind up together. But Carl would be here soon to drive with Elaine through the lighter evening traffic to the Dorset port of Weymouth.

'I'll stay to see you off,' she said, 'and then I must go. I've got a million things to do tonight.' She was planning to ring for a cab, but Euan insisted he would drive her back.

Carl arrived then and began to carry Elaine's cases to the car. He looked relaxed and happy, and Jenny could see Elaine blossom in his presence.

'We'll follow the holiday-makers as far as the junction,' Euan said, smiling . . . but then his smile faded and Jenny knew he was remembering that other procession, one car following another and Jed rising Neptune-like from the Rover's sun-roof.

'Bye-bye,' Elaine said, when the car was loaded. 'I'll bring you some wonderful scent back, and a bunch of pink lilies.'

'Take care,' Jenny said, kissing her and holding out a hand to Carl.

She climbed into Euan's passenger-seat as the other car drew away from the kerb.

'This was a *good* idea,' Euan said. 'Elaine'll come back a new woman.'

They halted at the junction, watching the traffic speed by on the Finchley Road. It was fairly light by London standards, but then it was eight o'clock and the home-going rush was over.

Jenny saw the lights change and Elaine's hand coming through the side-window, waving farewell. Carl's car moved forward, and Euan was easing his handbrake to follow, when Jenny heard the screeching of brakes, the terrible squealing of tyres. A red sports car, bursting through the lights in its haste to get down into the centre of London, caught Carl's car broadside on, sending it spinning wildly into the centre of the junction until it crashed into a lamp standard.

'Oh no!' Euan shouted. Jenny heard a voice beseeching God and then realized it was her own. There were horns blowing and the sound of rending metal, and suddenly someone was screaming, there was an explosion, and smoke and flames were leaping into the air as the traffic all around built up into a hopeless tangle.

'Come here, Jenny,' Euan said and pulled her to him. 'There's nothing we can do.' She was about to upbraid him for callousness until she felt the silent sobs that were shaking his chest.

'Are they dead?' The stupid question issued from her lips but she couldn't hold it back.

They climbed out of the car and ran towards the confusion.

'Get back . . . get back there . . . move on!' A traffic policeman was shouldering them away, pulling off his gauntleted gloves, bending his head to a walkie-talkie attached to a strap.

'A fatal . . . Finchley Road, junction with Albemarle Road . . . The lot, double quick. There's bloody chaos here . . .'

'Move along, *please*,' he said as Euan and Jenny came closer.

'We're their friends,' Euan said quietly. 'Is there anything . . .?' There was a stench of burning in the air, a dreadful heat from the flames, and somewhere a woman was laughing hysterically.

'We'll have lifting gear here shortly, sir. If you'll just hold on . . . I'll need details.' He was carefully non-committal, but there was no mistaking his meaning.

'Elaine is dead,' Jenny thought. And said it over and over again in her head as if repetition was the only way to force it home. '*I still can't believe it*,' Elaine had said. '*I can't believe I'm going to Jersey*.' Now she never would.

'I ought to cry,' Jenny thought, 'I ought to fall apart.' But she felt nothing at all. Only her teeth began to chatter and went on, getting wilder and wilder, until Euan, white-faced, had to shake her to make them stop.

'I saw your lad on the telly, Steven,' the man said as he entered the club. 'When are we going to see him in here, giving a turn?'

'He's not a bloody ventriloquist,' Steven said, 'he's a presenter.'

'I know that,' the man said. 'We're not all ignorant oiks because we didn't go to Oxford University, you know. I went to public school, an' all.'

295

There was a roar of laughter. 'Granny Pit,' one wag shouted, 'that's where you got eddycated, Leo.'

'Neewhere better,' the man retorted.

Steven left them to their chatter and moved into the back room. He had watched Keir that afternoon, playing a daft quiz game with forfeits. Granted he'd got most of the answers, but they hadn't been high-class questions, not by a long chalk.

He carried his pint to the corner and sat down. Surely Keir would take a break and come north soon? Cissie didn't say much, but he knew she hankered for the lad. Not that he was a lad now: he was twenty-eight next birthday. Perhaps he'd come home for that.

There was no one else in the quiet room and, in spite of the noise, only a handful in the bar outside. The strike must be over soon: it was only the need for an honourable settlement that was holding things up. The fight had gone out of most of the men, and no wonder. Everywhere you looked faces were grey, especially those of the women who bore the brunt of it. Except that with men it was a matter of pride: you provided for your family. Durham men had always done that, until now.

And it wasn't just men and women – yesterday he had heard a kid crying for crisps. Real tears of misery. He had fished in his pocket for coppers, but the mother had slayed him with a glance.

'She's got to learn,' she had said and hurried the child away. And a moment later Steven had seen a dog, ribs sticking through its skin, scavenging a dustbin.

He lifted his glass but the beer tasted strange now, unwelcome on his lips. Life was a bugger, and no mistake. He forced himself to drink it and then went out into the street, ignoring the good-natured ribbing as he passed through the bar. As he walked home he counted the graffiti: *'Police are Pigs'* on one fence: *'Kill Scabs'* on another. My God, it was a bloody war! And then he saw the war memorial, with its dates of 1914–1918 and 1939–1945. Under them someone had sprayed: *'The People's War 1984–5.'*

Steven tut-tutted at the sacrilege but it was all you could expect nowadays: there was no respect any more, not even for parents. 'You're turning into a sour old bugger,' he told himself – and then he remembered that funny little kid that was staying with Botcherby's lass. 'Don't nag,' she had told Cissie this morning when she went on at him about his dirty boots. 'Don't nag' – just like that! For all the world like a granny. It was coming to something when you had to be stuck up for by a bairn. All the same, she was a nice little thing, and Cissie could bring her back whenever she liked.

One day, if Keir got himself sorted, they might have grandbairns but in the mean time the funny little thing with her beads and her varnished nails would do.

27

2 February 1985

Kath lay for a few moments, savouring the fact that it was Saturday. No picketing, no rows, no screeching police cars. Peace, perfect peace. Dave would be home all day to help with the kids; she had food in for the weekend. One way and another she felt good – until she remembered the time-bomb ticking away inside her. Oh God, she didn't want to be pregnant, not now.

Dave was turning towards her, sliding his arm around her, too drowsy to be a threat.

'OK?'

'Yes,' she said, putting up a hand to pull the sheet up around his shoulders. 'Go back to sleep.' In the next room she could hear faint stirrings, Brian giggling and L.J.'s voice, imperious even at a distance. She would have to get up before they got into their stride.

'Is mammy coming today?' L.J. said when Kath entered the bedroom. Brian was kneeling up in bed, pyjamas rumpled so that the Batman motif was almost indecipherable. She held out her arms to her son, lifting him to sit astride her hip. He was four years old, but still her baby. In the big cot Margaret slept on – poor little Margaret, who would be elbowed aside if there was a new baby, already turned out of her bed for L.J.

Suddenly she was aware of movement. L.J. was coming nearer to her looking woebegone.

'Come and get a cuddle,' Kath said, 'and then we'll have some breakfast, and then we'll talk about your mammy and when she'll be coming home.'

'I'll tell you what,' she said as she shepherded boy and girl downstairs. 'After breakfast we'll make a calendar and tick off the days till mammy comes.'

Later, she watched Dave and the kids until they were out of sight, Margaret in the pushchair, Brian and L.J. walking on either side. They were going to the park and she would have peace.

Elaine would have to do something about calling L.J. by a civilized name before she went to school – and that would be soon, Kath thought as she moved to the sink. The child would be five in August: 'On my wedding anniversary,' Kath remembered. She had been so happy that day, so overwhelmingly bloody certain.

She turned on the tap and started on the vegetables. L.J. was a good little eater, thank God. Anything and everything that came her way. She had even perked up Brian's appetite by clearing his plate for him if he didn't eat up quick. Kath found she was smiling to herself. If this *was* a baby, she would cope. The strike must end soon and there would be good money to be made when the pit reopened. 'We'll have to work Saturday and Sundays,' Dave had said. 'There'll be a lot of deterioration to take care of before we can be in production.'

She was humming along to the radio when she heard the car at the door, and then the discreet knock. It couldn't be somebody dunning for money, not on a Saturday lunchtime. She opened the door. Jenny's father, Mr Sissons, was standing looking sombre on her doorstep.

'Can I come in, Kath? I'm afraid I've got some bad news.'

'It's not Jenny, is it?' It couldn't be, mustn't be Jenny. Kath felt a sudden weight on her chest, a bearing down inside her gut.

'It isn't Jenny,' he said and moved past her into the kitchen. He looked around. 'Are the children out?'

'Yes, they've gone for a walk. Dave's got them.' Kath was feeling behind her for a chair, running her hand along the back and down, feeling for the seat before her legs gave way.

'I'm afraid it's Elaine. And her man-friend . . . Carl, I believe?'

'They're not *dead*?'

'I'm afraid so. A tragic road accident. Jenny was there and saw it, so there's no possibility of a mistake. She

wanted you to know, straight away, for friendship's sake –
and because of the child.'

'My God,' Kath said, sitting down on the upright chair.
'That poor, poor little girl!' She knew she should offer
Jenny's father some tea or at least some courtesy, but she
couldn't seem to function. All she could think of was the
string of pearls she had just fastened around L.J.'s neck.
'These are my mammy's, aren't they?' the child had said.
'I'm keeping them till she comes home.'

There had been two messages from Annie on Keir's
answerphone when he got in at two. He had cleared them
both and gone to bed. He couldn't avoid her for ever, but
he didn't intend to speed the evil moment – let her cool off
first. She had been told about the changes yesterday
afternoon, changes effective from that day. As of now Keir
was executive producer of his own show, with the power
to choose both producer and director.

He slept well and woke in time to hear the nine o'clock
news. Nothing earth-shattering. He ate fresh figs for his
breakfast and carried coffee to the window seat with its
view over London rooftops. If he stayed here he would
have the place done over. Get one of the top names, or
perhaps dream up a scheme himself. He had been clever at
design once, and it would look good in profiles: '*In his
private life Keir Lockyer spends hours in his studio, creating
schemes for modern décor.*'

Except that there weren't enough hours in the day. He
would get someone in. He could ask who was hot at the
moment, that would make sure it got into the columns. He
lit one of his two cigarettes of the day and leaned back
against the window frame. Perhaps it was all too good to
be true, and he would wake to find it was really Saturday
morning in Belgate, with mam frying up and dad chunter-
ing on about the Party? His father had always been
ambitious for him, but he had surely excelled those
ambitions now? He would be bigger than Frost, or Wogan
even, but only if he got the mix right, that blend of
topicality and entertainment that kept the punters glued to
the screen.

300

He was still considering the right mix when he heard a key in the door. 'Shit!' He must get that bloody key back, or change the locks.

'Keir.' Annie was smiling and she looked good, in grey slacks and matching jacket slung round her shoulders over a white silk shirt. Classy. He felt himself warm to her. The lady had guts.

'Annie. I was just going to ring you.'

'How kind.' She was detaching the key from her ring and laying it on the hall table. 'I won't trouble you long. There are one or two things of mine . . . it won't take a moment.'

He followed her into the bedroom, watched her take her spare pills from the bedside table, her robe and slippers from the ottoman, the little Lalique clock from the bedside table. Damn, he had liked that clock.

'Annie, I heard about your move. Exciting for you. Africa . . . the Dark Continent! And they say you'll have a free hand.'

'And no budget,' she said. 'Just what I like, a challenge.'

'I hope you realize it was nothing to do with me.'

'Don't worry, Keir, I know precisely what part you played in it.' She was putting up a hand to the watercolour above the fireplace but then she suddenly withdrew it. 'I'll leave you that. I never liked it anyway.'

She produced a cheque book and pen from her bag. 'Fifty pounds, Keir, for any phone calls I may have made this quarter, bath water, et ceteras. I always pay my debts. Always.' It was a threat and he knew it. More than a threat, it was a vow. Annie was smiling suddenly. 'If you want a breakdown, I'd say twenty pounds for the calls, twenty for the bath water . . .' She was raking his body with her eyes, dwelling on his crotch for a second, and then meeting his gaze. 'The remaining ten is for the et ceteras. They didn't amount to much.'

Keir felt colour flame in his face as she swept out of the flat. And then he remembered why she was so bitter. Let the bitch carp – he could afford it.

*

301

Jenny had done everything she could think of doing: contacted her parents and asked them to break the news to Kath; left a message on Alan's answerphone; phoned the Durham police and given them such details as she could of Elaine's parents. 'Can you ask them to ring me, if you do manage to locate them? They'll have to make the arrangements since they're the next of kin. I'm not trying to duck out of it, I just don't have the authority to act . . . and someone must.'

The police had been courteous but non-committal. 'It's not a lot to go on, Miss Sissons, but we'll try. You're sure the name was Gentry?'

'Yes, Elaine's name was definitely Gentry – but her stepfather will be called something else, and I'm afraid I don't know what.'

'We never expected to die,' she wanted to say. 'We none of us thought we would only meet again at funerals. We were the future, that's what we thought. That's what Jed said.' She started crying before she put down the phone and afterwards she couldn't stop.

She was still crying when the doorbell rang and as she opened it, wiping her eyes, Alan was stepping over the threshold, removing his coat with one hand, reaching for a hanky with the other.

'Come on, Jenny, I know how you feel.' It was comfortable to be in his arms, to feel the buttons of his waistcoat against her cheek just as her father's buttons had both irritated and comforted her when she was a child.

'I'm glad I was in London,' he said. 'Eleanor was first back. She got your message and rang me at the office. I was catching up there . . . nothing that can't wait till Monday.'

So he and Eleanor were living together, Jenny thought. Well, it was only to be expected . . . except that she hadn't expected it. Keir, yes, with his TV executive, but not Alan. Only she and Euan were left alone now. She started to cry again and Alan shook her gently.

'Shut up, Jen. Where do you keep the booze? We could both do with a drink.' Her teeth had started to chatter again in spite of the fact that she felt hot and agitated and her face was swollen up and sore. She showed him the

302

drinks cupboard and bent to turn the fire up, then took the glass he offered. It clinked against her teeth. 'We're jinxed, Alan, I can feel it. We're none of us meant to be happy. Jed. Elaine. Kath with so much trouble on her plate. Euan is devastated by it all.'

Alan took her glass from her and gripped her by the shoulders again.

'And you and I were perfectly happy until this happened, Jenny. Get a grip on yourself. What's happened is tragic, but it's not malign fate or any such thing – it's life, and we can't escape it. Now, we have to make plans, not least for Elaine's daughter, so drink up and let's get on with it.'

Jenny took the glass again meekly and let the spirit burn her throat, comforted to think that at last someone else was in charge.

Cissie's face, as she came in holding a carefree L.J. by the hand, told Steven that something was wrong.

'What is it?' he said as she took off L.J.'s anorak.

'Put the telly on, chuck,' she said to the little girl, helping her push the right button. 'There now, cartoons. Pull the cracket up. That's your little chair.'

'It's terrible, Steve,' she said as they went into the kitchen to get the child a drink. 'Kath just heard. It's her mam . . .' She gestured towards L.J. 'She's dead, killed in a road accident. I've just left Kath. Dave's taken their bairns to his mam, but he couldn't very well ask her to take three. I said I'd bring her here for as long as they wanted. It's all right, isn't it?'

'Why, of course it is. Poor little bugger. Where's the father?'

'No one knows,' Cissie said, rolling her eyes. 'You know what they're like nowadays. The girl was bringing the bairn up on her own – that's why Kath took her, to give the girl a break. And it's killed her. On her way to Jersey, she was, in a car with another feller. It was a stolen car that did it, young kids joy-riding. Joy-riding! She's dead, and the man that was with her, an' all.'

'Maybe he was the father?' Steven said.

'No, he wasn't, he was someone else. It's too complicated for you and me, we still play by the rules. Any road, we have a chance to have her here for a while so I'd best get some tea on. What do you fancy?'

'To tell you the truth, Cissie, I haven't got much appetite. Five years old . . . not five years old, just a babby . . . and she's got no one.'

'Kath won't see her left,' Cissie said. 'I don't know about Dave, but Kath's a good-hearted girl.'

'She'd need to be, to take a strange bairn on,' Steven said.

Cissie sighed. 'That's right, look on the bright side. Come on, let's get cracking. I've got a bit of black pudding and bacon. With a nice tomato?' She was coaxing him and he felt his saliva stir.

'Well, mebbe a bit. Don't make much.'

He went back to his chair and had hardly settled when the funny little thing got up from the cracket and settled herself on his knee.

'Get comfortable,' he said with heavy sarcasm.

'I am,' she said equably. 'Just move a little bit that way.' She turned and smiled at him. 'Please,' she said.

He moved, grumbling theatrically as he did so, resisting the impulse to take her in his arms and hold her tight.

'It's my fault,' Euan said. They had done everything they could think to do, and now they sat, Alan and Euan in the chairs either side of the fire, Jenny on a pouffe between them.

'Cut that out,' Alan said. 'No one could've done more for Elaine, so don't start wallowing in entirely misplaced guilt, Euan. Tell us what you learned from the police, if anything.'

'There wasn't much,' Euan said, turning his drink round and round in his hands as he spoke. 'Death was instantaneous for both of them. There'll be a post mortem, but the inspector said the hospital was sure.' His voice broke and he sipped his drink before he continued. 'The car was stolen, and the kids who took it didn't even have a licence. Of course, that means they weren't insured.'

The phone rang suddenly and Jenny went to answer it. After a little while they heard her profuse thanks, and then she came back to the pouffe.

'They've found Elaine's parents. Apparently they're going to be ringing me. The police gave them my number.'

'That's a relief,' Alan said. 'We'd have been in a dicey position without them. We have no right to act, but someone has to. Did they say anything about the inquest?'

'Just that there'd be one,' Euan said. He looked down at his drink. 'I wanted to . . . see Elaine but they said best not, for the time being.' When he looked up his eyes were agonized. 'She burned to death, Alan.'

'She died instantaneously – you just told us that.' Alan's voice was cold and factual, and it comforted Jenny. He knew what he was talking about.

'If I hadn't pushed her into having a holiday she'd be at home now,' Euan said. 'Bathing L.J., playing with her. They used to run around the place like two kids, and she was the worse . . .'

His voice broke and Jenny reached to touch him. 'Don't hurt yourself, Euan. Elaine's life would've been nothing this last few years without you, she told me that often. You did what you did about the holiday out of love. That can't be wrong.'

The phone rang again and Jenny went to take it. This time she stood quietly, obviously listening to a long tirade. When at last she put down the phone her eyes looked dark in a white face.

'Elaine's parents don't want anything to do with it. They'll sign anything, agree to anything . . . as long as we handle everything.'

'That's fine,' Euan said. 'That's the way we want it.'

'What about the child?' Alan said. 'Long term, I mean? Aren't her grandparents the natural people to take her?'

Jenny shook her head. 'Elaine wouldn't want them to have her. She didn't get on with them. No, L.J. can't go to them.'

'Let's take it one step at a time,' Alan said. 'First, we have to make the funeral arrangements.' They doled out jobs, each taking a share.

'I'll write to Barbara,' Jenny said. 'Or perhaps I'll phone.'

'Has anyone told Keir?' Alan asked.

Jenny bit her lip. 'No,' she said, 'it never occurred to me . . . But we should tell him.' But should he also now be told about L.J.? Should anyone else know the truth? Had she a right to keep it to herself?

'Go to bed, Jenny,' Alan said, taking her glass. 'I think you've had enough for one day.'

28

8 February 1985

Kath slept fitfully, waking every time the bus pulled up or slowed down. She had never travelled overnight on a coach before, not even in her student days. Sometimes she peered from the window, hoping for landmarks, but the motorway was black and featureless.

She had sold her father's half-sovereign to raise the fare: the Pakistani owner of the corner shop had given her £60 for it. Both Jenny and Euan had offered to pay her expenses, but Euan had already been generous enough and besides, Kath wanted to pay her own last tribute to Elaine. Increasingly she could see that Elaine had made a better job of rearing her daughter than anyone had given her credit for. The child was generous and kindly, protective of younger children, and open to reason even when she made her many demands.

'I'm really fond of her,' Kath thought and wondered for the dozenth time what would happen to L.J. Dave had agreed to keeping her until proper arrangements could be made, but it couldn't go on forever. Soon she must tell him she was pregnant, and that would make a difference, whether or not she went ahead with the pregnancy. She must tell her parents, too, which was something she really shirked.

She eased her position against the window, checking her watch. Another three hours before she got to Victoria. Jenny had told her to take a taxi to the flat. 'I'll have the kettle on and breakfast ready, so be as quick as you can.'

Jenny was seeing to flowers from her and Dave and the kids, and a separate wreath from L.J. They had deliberated on the phone for a long time before deciding that L.J.'s card should simply say, 'I love you, mammy' and her name. They had both cried then, and afterwards Kath had gone back home from the call-box and looked down on the sleeping child, her mother's beads on the table beside her bed.

It was a good job you couldn't see what life had in store for you, she thought, remembering the hill and their high hopes. How many of them would have done things differently if they had foreseen their future? Jed and Elaine, certainly. As for herself, would she do it all again or not? She thought of the unpaid bills, the mortgage arrears, the gaps in her children's clothes. At the moment she and Dave never saw their creditors; not even tallymen expected to get blood out of a stone. But as soon as the miners went back to the pit they would all be there, demanding their pound of flesh. And they couldn't all be paid at once. There would not be enough to go around.

Kath must have dozed then, for when she woke it was grey daylight and her bus was one of several disgorging into a bleak and windy Victoria bus station. She took a taxi to Jenny's address, and was surprised at the size of the house, winded by the three flights of stairs she had to climb inside.

Jenny had been as good as her word: there was bacon and sausage, toast on a rack, and as much tea as she could drink.

'Heavenly,' Kath said and then, desperate to tell someone, 'I'm pregnant again, Jen. Another mistake. No one knows but you because I'm not sure I can go ahead with it this time.'

'Oh Kath,' Jenny said, sitting down opposite her. 'It's not because of the strike, is it? You mustn't let that influence you. Paul says it won't last out the week now, and he knows about these things. And I have some money put away – it's yours whenever you want it. Euan would lend, too. You really mustn't shut us out if we can help.'

'Ta, Jen. I'm grateful. And if it was just money I'd take a loan like a shot. But it's more than that: it's do we want another child, at this time or any other? And the shortage of money won't be a temporary thing, this has set the miners back a decade. Sometimes I think we'll never get on our feet again.'

Alan came to join them early. 'Eleanor is going straight to the church,' he said, 'but she thought you might need me here.'

'That was kind of her,' Jenny said. 'She's coming back here later, isn't she?'

'I think so. But she does have a case conference . . .'

Kath came out of the bedroom, picking imaginary fluff from her grey suit.

'Hallo, Alan . . .' The words 'it's lovely to see you' formed on her lips but were never uttered. Instead they moved to one another and embraced.

Euan arrived next and they all clung to one another without speaking. Then the four of them stood awkwardly around the fire until Jenny decreed a drink. It was good to have something in their hands, but no one drank with gusto. Euan's fingers were visibly trembling.

'Do you remember how Jed used to fold her in that awful green coat of his?' Alan said suddenly, and they all felt better now that emotion was out in the open.

'He thought she was the bee's knees,' Jenny said.

'Well, so she was,' Kath said. 'She stuck by her kid . . .'

'And she never grumbled,' Euan said. 'I never heard her say "Why me?", and I seem to say that a lot.'

'And she was always pleased with things,' Jenny said. 'Even little things and she'd jump for joy. She'd have loved Jersey, but she never got there.'

All the same, Jenny had done her best. . .she had thrown herself on the mercy of a West End florist, describing the pink lily she had seen in Jersey and how Elaine's eyes had lit up at the mention of it, telling the florist enough to enlist her sympathy. 'It'll be an Amaryllis belladonna,' she said thoughtfully, and the next day produced one perfect specimen. Whether or not it was the flower Jenny had seen in Jersey, it would do as a symbol. It rested now in Elaine's clasped hands, and the thought of its being there comforted Jenny as they left the house and took their seats in the leading car.

At the church they climbed out, surprised to see a gaggle of people at the gates.

'There's a camera over there,' Kath said. 'No, two.' And then Keir was beside them, his face grave, bending to kiss

309

Jenny and Kath, embracing Euan and shaking hands with Alan.

'It's awful, isn't it?' he said. 'I couldn't believe it. I should be doing a pre-record now but I said it must wait. Old friendships . . .' He raised his hands in the air and shrugged. 'Is Elaine's child all right? It's with you, Kath, isn't it? Mam told me when I phoned.'

The cameras were moving closer, snapping away. A girl, obviously a reporter, was scribbling in a notebook as Keir bent to speak to her.

'How did they know?' Kath asked, furious.

'Telepathy,' Alan said. 'Now form up and let's get inside.'

They sang the hymns they had sung for Jed because they had all agreed that Elaine would have wanted it so. The vicar spoke briefly of the tragic loss of a young mother, and of another life lost at the same time.

'We've all forgotten Carl,' Jenny thought. 'It's as though he didn't matter because he wasn't one of us,' and she made a mental note to write to his parents, as they formed up again behind the coffin, with only L.J.'s wreath upon it.

The ceremony in the crematorium was mercifully brief and then the rollers were carrying Elaine away from them forever. Jenny tried to think of her, so thin and waif-like, wiping her nose on the back of her hand when she cried, running barefoot away from the car that New Year's Day, carrying her rickety sandals in her hand, turning to call 'ta-rah'. 'I want to remember her,' she thought. 'I must feel the pain of her going now, so that I will never forget her.' But in her heart she knew she would remember Elaine because she belonged to youth, that period of life when the senses are camera-sharp, taking pictures that are recorded forever, to be played back at will.

The reporters were still waiting when they came out, and Keir dropped back to speak to them again, shrugging his white mac higher on his shoulders, shaking his blond head as though in disbelief.

'Do you think he's turned into a nancy boy?' Kath whispered. 'I know what Dave would say if he saw him.'

'If half the stories I've heard are true,' Jenny whispered back, 'that's about the one thing you can't accuse him of.'

In Jenny's flat they nibbled on titbits and drank a lot. Eleanor never appeared, but Alan seemed not to notice. Jenny wished Paul had been there with her but they had decided not to risk it and after seeing the cameras at the church Jenny was extremely thankful.

Had Keir arranged the publicity? Hard to believe that, looking at him sitting on the floor, his back against Kath's chair.

'Oh God, it's good to see you all again,' he was saying. 'If only . . .' And then he was reminding them of the past, telling stories as only he could tell them.

'. . . so old Matthews gives us this long spiel about planting a tree wherever and whenever you can. Jed waits till break, and then he nips off, pulls up a young sycamore and plants it in the staff bog. 'He wants trees, he'll get trees,' he says. It was drooping all over the place and they knew it was him, but they daren't come out with it.'

'We all got kept in,' Kath said ruefully, 'I do remember that. Still, he was worth it.'

'I wish I didn't have to leave,' Keir said, after a last whisky. 'I feel wretched that it's taken something like this to bring us together. I'll be in touch soon. Soonest.' He pressed Jenny's hands and kissed her brow. 'This week.'

They called their goodbyes and bandied promises to ring, but they were all relieved when he was gone.

'Let's face it,' Kath said, 'he's not one of us any more. It's not just that he's successful – you're all of you successful in one way or another . . .' She grinned at Euan. 'Or rich, which amounts to the same thing. But I still *know* you. I could count on you if I needed you. He's like quicksilver, now – press him, and he slides from under your fingers.'

Jenny was remembering the pressure of Keir's fingers on her palms. Once upon a time that would have set her aquiver. Now she had Paul, and that made a difference.

'I'll keep L.J. with us for the foreseeable future,' Kath said, when they had slumped into chairs with mugs of coffee. 'I've talked to Euan and accepted his offer to pay for her because Dave and I are skint. I can't promise how long I'll have her for, because things are a bit rocky back home

311

at the moment. But until it's decided what's best, she's safe with us.'

'Dave's out picketing in the morn, so the bairn's stopping here,' Cissie had said last night when she brought L.J. to stay. 'His mam's got the other two stopping. Kath won't be back till tomorrow morning, but Dave won't be out as it's Saturday, so it's just the one night.'

'All right, all right, I don't need convincing,' Steven had said. He was teaching L.J. a rudimentary form of dominoes and she was proving an apt pupil. They let her stay up until her head drooped on Steven's shoulder: 'You can't be hard on a bairn in a strange house,' Cissie said. They were forbidden to tell her anything about her mother's death – not that either of them was tempted.

'I wouldn't know how to tell a bairn,' Cissie reflected, and Steven shook his head.

'It could affect a bairn like that, a clever bairn. Even turn her funny.' For L.J. was clever – he could remember the signs from when their Keir was little. This one had a lot of the same ways of putting things, of jumping ahead of you just like Keir all those years ago. They told him about the bairn on the phone, and he was pleased they were lending a hand.

'He was never close to the girl,' Cissie said, when she came off the phone. 'But they were all in the same crowd. You know what young'uns are like.'

Next morning L.J. woke and came into their bed, climbing across him and pushing down between them.

'Move along,' she said to him, for all the world as if she owned the place. Steven grumbled, and then moved just the same.

'Now I'll have a story,' she said.

'Steven?' Cissie said.

'I don't know any stories.'

'You do,' L.J. said. 'You tell fibs.'

'So he does,' Cissie agreed, snuggling down.

'Two on to one,' Steven growled. 'I'm not playing.' But he recited 'Round and round the garden went the teddy-bear' and 'Itsy-witsy spider', and pulled his trousers on over his pyjamas to carry her downstairs on his shoulders.

'You treat her like royalty,' Cissie said, pretending to be huffed. Steven didn't argue. How else could you treat a bairn whose mother was being buried that same day?

The men took Kath to Victoria when it was time for her coach. 'I'll be up soon,' Jenny promised, thinking of the job they must do together when she went north. 'And remember what I said earlier, about not letting anything influence you?'

'Yes,' Kath said, 'I'll remember. And ta, Jen.'

'Will you be OK?' Euan asked anxiously, pressing chocolate and cans of coke into Kath's holdall.

'I'll be fine. It's a lot easier dozing on a bus than looking after three under-fives.'

'I can imagine,' Alan said, wryly.

'Look after Jenny, both of you,' Kath said. 'And don't forget where I live. You'd get on all right with Dave once you got to know him.'

'He must've had a tough year,' Alan said.

'Yes, he has. But it hasn't broken him, thank God.'

'And now the strike's as good as over,' Euan said.

He and Alan stood together, tall and handsome, so that Kath was proud of them both. They had not been a patch on Keir in the old days, not where looks were concerned, but Euan had improved with age. He looked graver, now, and the red hair was paler and more subdued. As for Alan, the old diffidence was gone, and his face, a little heavier with age, was positively handsome. 'But I wouldn't like to cross him,' Kath thought as he raised a hand in farewell and the bus gathered speed.

She thought about Belgate, then, about how they would be managing without her. She had certainly missed them – but it had been such fun to talk again, to argue about politics and morals and everything else under the sun. 'I'd forgotten how much fun it was,' she thought and was suddenly consumed with guilt at the betrayal of the man who was waiting for her even now, minding three children so that she could go to London to stand alongside old friends.

*

313

Jenny cried for a while when they had all gone, boozy snivels of misery. Life was cruel. She had hated it in the church, with none of Elaine's family there and Keir posturing away outside. Damn Keir. Why wasn't he the kind of father who could take proper responsibility for a child? Care for others, like most people did? How different things would be if she could tell him about L.J.

As it was, where would L.J. go? They couldn't put it all on Kath, forever. And she herself couldn't cope with a child, not unless she left the Beeb. Euan was a possibility: he could afford it, and he would never see L.J. go into care. But could he cope with her, bring her up? Jenny poured herself another drink and curled up on the settee, the weekend looming before her like a waste-land. Three days before she would see Paul: seventy-two hours. At least she could talk to him then, even tell him the truth about L.J.'s parentage. He would know what she should do . . . after all, he had children of his own. She cried harder, then, and only ceased when the doorbell rang.

She looked at the clock: half-past ten. It must be Euan, unwilling to face his empty flat. She could put the cushions from the settee on the floor and he could sleep here tonight, then neither of them would be lonely.

But it was Paul who stood on her step, his case in his hand.

'I got to Euston, Jenny, and I couldn't go – not this weekend. I'll stay with you tonight and go down in the morning. I rang Rose and made some excuse. She'll be OK.'

Jenny took his coat and poured him a drink, and then let him wipe her face with his hanky and smooth her hair from her brow as her mother used to do.

'I love you, Jenny. God help me but I do.' She went into his arms, revelling in the comfort she found there, knowing that what was happening between them could no longer be controlled. It would shape itself, for better or worse, and all she could do was take from it whatever joy was to be found.

Later, as they sat by the fire, arms linked, she broached the subject.

314

'We've been seeing a lot of one another, and you know how much I like . . . love . . . that.'

He put out a hand to press hers. 'But . . .?'

'But I don't want to do or be part of anything that creates difficulties for you.'

'Let me worry about that, Jenny.'

'I can't, Paul, not entirely. I care very much what happens to you, and if I thought I was going to damage your career . . .'

'Now that you can leave *entirely* to me. I'm too outspoken to get promotion, Jenny – even in Opposition. So if you're worried about nipping my preferment in the bud, you can forget it.' Suddenly he was serious. 'I worry too – about you. I don't know at this stage what, if anything, I can offer you. Am I queering your pitch? That's the real issue.'

Jenny shook her head. 'If you mean are you deterring my string of eligible suitors, you're not.'

'So we go on as we are for a while?'

'So we go on as we are.'

'Good.' He was smiling as he spoke.

'I've got heaps and heaps to do,' Jenny said. 'There's been so much to settle since Elaine died, and my own affairs have got in a bit of a muddle . . .'

'Have you sorted out the future of the child yet?' Paul asked.

'No, she's still with Kath but that can only be temporary.' She felt panic rise in her chest so that she gasped a little as she poured out her fears. 'I can't let her go into care, Paul, but how can I take her? I've gone over and over it. I'm the logical one – female, affluent and unencumbered. But how could I work *and* cope? Apart from me there's only Kath, and she has her hands full.'

They talked then, arguing the pros and cons, and although no solution emerged Jenny felt her tension ease.

'Euan is fretting too,' she said at last. 'If he were to apply for custody of L.J. would he get it, bearing in mind that he's a single man?'

'I doubt it. Control perhaps . . . he foots the bill, she's fostered, but he makes decisions in education, et cetera. That might be a possibility. But would Kath consent to keep her on those terms?'

315

'I doubt that she could. And we have to consider Dave –
it's his home, too.'

'Well,' Paul said at last, 'try not to think about it any
more now. You look so tired. Let's talk about holidays. I've
been wondering if we could get away together for a
while . . .'

29

15 February 1985

Jenny's desk was cluttered with cuttings and briefing sheets and the cascade of spring flowers Paul had sent her for Valentine's Day. She reached out to press down a stray iris and then went back to her difficult decision: did she do contraception – a hot item since the ruling forbidding doctors to give contraceptive advice to girls under the age of sixteen – or the question of a civil servant's loyalty, which was in the news since Clive Ponting's acquittal by a jury on Monday? Yesterday he had resigned from the Ministry of Defence, whose secrets he had earlier revealed to a Labour MP.

Jenny pushed back her chair at last and walked to the window, not for the restricted view but because her work had become oppressive to her since Elaine's death. She was tired of struggling to achieve balance, fed up with protocol, above all sick of hearing people pontificate. If she did contraception harridans would be screeching on either side; if she did Civil Service loyalty, axe-grinding MPs would be out in force! Sod them all. She could always do Aids, that mysterious and worrying new spectre they must all come to grips with soon. But she was still on probation, and young to be producing in current affairs. If she got stroppy or over-controversial she would be out, and once out in broadcasting you were inclined to stay out.

She went back to her desk, wondering once again if she had chosen the right job. In the media you could come up fast, like Keir, and fall even faster, like Simon Dee. 'If I don't get a move on now I won't have a problem . . . or a job,' Jenny thought and plumped for contraception. Whether or not Mr Ponting was in the right, she didn't like the look of his unctuous face, and not having to meet him was one tiny privilege she could afford herself.

She reached for the phone and began to ring the names on the contraception contacts sheet. With a bit of luck she'd have it all lined up by lunch-time, they could have the planning meeting this afternoon, and she'd get home on time for once. It was Friday so there would be no Paul, but she had lots of jobs to catch up on around the flat and a letter to write to Kath, outlining her plans for a journey north.

At the other end of the line a lack-lustre voice recited a number by way of greeting.

'Oh, hello. This is Jenny Sissons here – I'm a producer working on "Public Enquiry". We're hoping to discuss the question of contraception in our next programme and I wondered how you'd feel about taking part?'

The voice at the other end was suddenly transformed, vibrating with life and eagerness to be heard on networked radio.

'Why do they do it?' Jenny thought, for the umpteenth time. 'When we use them we restrict them, cut them short, edit out their favourite lines, and give them a glass of inferior wine and a minute cheque by way of thanks. And they queue up to be exploited!'

It was a relief when light began to filter through the curtains and Kath could ease herself out of the bed. Dave slept on, sighing sometimes like an old man, but sleeping nevertheless. He had lain awake nearly as long as she, both of them staring down the dark wondering how they would cope, but he had only been worrying about feeding two children, whereas she had worried about feeding the child in her belly, too – and perhaps a fourth child, L.J. How could they turn her out? Who else could take her?

Dave woke then and prepared to go picketing in an ice-bound world. When he lifted the curtain to peer out there were frost patterns on the panes, and Kath could see a morose and solitary bird perched on next door's gable end.

Suddenly a song her father sometimes sang came into

318

her mind, written for some long-ago strike but still appropriate today.

> The flour barrel is empty now, their true and trusted friend,
> Which makes the miners wish today the strike was at an end.

The flour barrel was well and truly empty. Without the money Euan sent for L.J. they would be starving. She went down to make Dave tea and listen to the radio spilling news of men returning, defeated, to pits up and down the country.

Dave listened too, grimacing at the taste of tea without sugar. 'It's over really, isn't it?' he said.

Kath nodded. 'I suppose so.'

'And all for noth . . .' The word ended in a gulp and suddenly he was crying. Kath flung aside the teapot and took him in her arms, as though he was another, fifth child she must comfort and care for.

'How are we going to manage, Kath? How will we *ever* get on top of everything?'

'We will,' she said firmly. 'You get back to work and leave the rest to me. I've got a brain, haven't I? Well, now I'm bloody well going to use it. Give me a year – one year, Dave Gates – and we'll have money in the bank!'

'And a bairn in the pram,' she thought despairingly but forbore to say so.

'Have we fixed the Astons?' Keir demanded.

The girl researcher flinched. 'Not yet.'

'Well, put a bullet up Graphic's arse. Now!' He turned as she scuttled off. 'I want monitors everywhere . . . I want to be able to frigging well see what we're putting out, no matter where I am. Do you hear that?'

'We've got all the monitors placed,' the floor assistant said.

'Get more.'

'What does he suggest we do? Breed them?'

Keir heard the aside and grinned. 'Funny, funny. If you

have to fucking breed them, breed them, but get them just the same.'

He went on shouting orders till everyone was looking suitably alert.

'Right,' he said. 'Now I, on whose shoulders this whole incompetent edifice rests, will attempt to have a couple of hours' rest before the promo. Thank you, ladies and gentlemen. See you all at six sharp.' He would have liked it if someone had slapped him on the shoulder and wished him well with his first show as exec., but no one did. Because they were an envious bunch, probably. In any event, it didn't matter.

Keir looked at his watch: twelve-thirty. He ought to eat now and then get a couple of hours' kip, but he felt like unwinding. In the old days Annie used to keep him amused before the show, always coming up with something to divert him. Perhaps he could ring Jenny? She had looked elegant at the funeral. Quite striking. He had never thought her anything but pretty in the old days, but now she had pzazz. She wasn't far away at B.H. – they could meet for lunch and then he could go back to the flat, and she might come with him. She had been keen enough in the old days . . . that night in the shop doorway, when she had been to see Elaine . . . poor little Elaine. A born loser.

Perhaps, though, today was not the time to start something with Jenny. Afterwards, after tonight . . . He looked around his dressing-room, checking his appearance-clothes shrouded in a cover, the bottle of Niersteiner ready for the ice-bucket, the buckled patent shoes newly shined by Wardrobe. It was all OK. He swung his coat on to his shoulders and made for the exit, hoping the people waiting there in anticipation of a celebrity would not hold him up too long.

'Where are we going?' L.J. pulled on his hand to ensure an answer.

'To see your Auntie Cissie,' Steven said. They turned in at the church gates and made for the door to the hall. Noise and warm air gushed out at them and Steven swung the child into his arms.

'Watch out for Cissie, see if you can see her?' They found her in the back, straining a huge pan of taties, adding a half-pint of milk and a knob of margarine big as a fist. She pounded the mixture with a potato-masher, sending up swirls of mash, patting it all off occasionally into a herring-bone pattern.

'What do you two want?' She grinned up at the little girl in Steven's arms. 'Come for your dinner?' She had a whispered consultation with a stout woman with big gold loops in her ears and then directed them to a table. 'We've got plenty today, for a change, so you can eat here and save my legs.'

'What is it?' L.J. asked, seeing the steaming plates carried by.

'Kit-E-Kat stew, chuck,' a young miner said, waiting like them to be served. 'They say it's beef, but it's really pussy's pieces.'

'Watch your tongue, lad,' Steven said amicably. 'You'll put her off.'

'She'll eat up all right, won't you, pet?' The boy's T-shirt was frayed at the neck and his parka pockets were torn but Steven liked the look of him.

'Who are you, then?' he said.

'Baxter . . . Tony Baxter. Sammy Baxter's me dad. He was in the lamp-cabin. He knows you, Steve. . .why, everybody knows you on account of your Keir. When's he bringing his show to Belgate? I can give him a few opinions.'

'You never can tell,' Steven said. 'I know he's kept up with the strike all along – and given. There's more been given down his end than up here.'

'That's easy explained,' the boy said. 'We've got nowt, that's why that is.'

'You'll soon be back to work,' Steven said.

'Not me, mate. I'm a naughty boy. Two convictions for violence. I faint at the sight of blood – but I've beaten people up twice, according to them. They'll not have me back.'

'What'll you do, then?' Cissie was putting plates before them all and the boy looked down at the food.

'Same as I do each time I eat this lot . . . hope for a

321

bloody miracle. I wish I was retired like you, jammy bugger. Nee more tub-loading, nee more back-shift . . .'

Steven felt his mouth tremble. How could people be so blind? A job was life, the only thing that mattered.

'Nee more pay packets,' he said aloud. 'Thou go back to hewing when thou gets the chance. It keeps the wolf from the door.'

Kath had scanned the pages of a week-old evening paper, which was being passed around. No one could afford their usual papers now, and she, for one, missed them like hell. When, if ever, they were not skint, she would take all the broadsheets and at least one tabloid for flavour. She was turning to the first page to begin again when she heard the knock on the door. Oh God, let it not be someone wanting money!

The woman on the step was a bottle blonde dressed in a tight-fitting black suit with jet bugle beads on the shoulders.

'Are you called Botcherby?' she said.

'I was,' Kath replied.

The woman turned, calling to the man who sat in a battered car at the gate. 'It's here, Fred.'

Kath felt her hackles rise as the man lumbered out of the car and up the yard. 'Do you mind telling me who *you* are?'

'Our name's Foster and you've got my grandchild here.'

It took a moment for Kath to grasp what the woman had said and then she stepped back from the door. 'You'd better come in,' she said, glad that the kids were out with Dave but wishing he was here with her all the same.

'Where is she?' the woman said, looking round the empty kitchen. She was trying to sound grandmotherly and sweet, but it wasn't working and the man obviously knew it.

'We've got a right to see her,' he stated.

'No one's said you haven't,' Kath said. 'But we were told you wanted nothing to do with things . . . Elaine's funeral, anything.'

'Well,' the woman said, 'it was a shock. I mean, hearing it like that on the phone. And after all the time Elaine

322

hadn't been in touch. I just said the first thing that came into me mind.'

'And then we started thinking.' This was the man. 'About what was proper, and everything.'

'Well, you'd better sit down,' Kath said. 'I'll put the kettle on.' She had just enough tea-bags to see her through to the weekend, but this was an emergency.

'How did the little girl come to *you*?' the woman asked while they waited for the kettle to boil. 'With Elaine being down south, and everything?'

'We were friends,' Kath said. 'All of us in a crowd. I'm the only married one, so it had to be me who looked after her while Elaine went on holiday.'

'We thought it might be on account of Keir Lockyer,' the man said a little too nonchalantly. 'He comes from round here. In fact, that's how we found you – we went to the corner shop and asked the black chap. He sent us to Lockyers and they sent us here.' He looked round the kitchen and she guessed he had expected something more palatial.

'How did you know Elaine knew Keir?' Kath asked, scalding the tea in the pot.

'We saw it in the paper . . . him so upset and everything.' Too late the man flashed the woman a warning glance.

'I see,' Kath said, moving to the table with teapot and cups. So their interest stemmed from the fact that Elaine knew someone in the limelight, someone who just might have money. She would have to tread carefully now. It would be easy enough to dispel their fantasies about Keir, but Euan was a different matter.

As she poured the tea, Kath said off-handedly: 'Keir was friends with Elaine's boyfriend. Help yourself to sugar. Elaine's boyfriend was killed in an accident before L.J. was born.'

'Eljay?'

'Her name's Lavinia Jane. We call her L.J.'

'Was it a car accident?'

'No,' Kath said, 'he drowned . . . so there was no question of compensation.' They had assumed Jed was L.J.'s father: best to let them keep on thinking that. 'Elaine's had a hard time over the last few years. She worked in a pub and she's had hand-outs from friends. Not from Keir Lockyer, though; they were never that close. He just came to the funeral for old times' sake.'

'What's she like?' the woman said. 'The kid?' For the first time Kath saw a gleam of concern in her eye.

'She's a very nice little girl. A bit headstrong . . .'

'Like her mother,' the man said unpleasantly and Kath felt her resolve harden. L.J. was not going to them while she had breath. On the other hand, did she have the right to refuse them?

Childish chatter sounded in the yard, and then Dave was entering the kitchen, Margaret in his arms, the others at his heels.

'This is Elaine's mother and stepfather,' Kath said. She bent to L.J. 'This is your grandma. Your mam's mam.' Suddenly she realized that L.J. still did not know of Elaine's death. She mustn't learn of it here, in this crowded kitchen. Kath straightened and turned to the man and woman. 'L.J.'s hoping her mammy will be back soon. We've told her it'll be a few days.' Her eyes locked with the woman's and she saw comprehension there.

'Well, I expect she will,' the woman said. She held out her hand to L.J. 'Come and see your grandma.' She was fishing in her handbag. 'I've got some sweeties here for you . . . here they are.'

But L.J. had shrunk back against Dave's legs.

'Come on,' the man said, 'say hello to your grandma. We've come to take you to our house.' L.J. did not move.

'What are we going to do?' the woman said to the man, her eyes fixed on his face.

Before he could answer Dave had passed the baby to Kath and was gathering L.J. up in her place.

'I'll tell you what you're going to do,' he said, smiling. 'You're going to get to hell out of my house, and if you want to come again we'll need notice.'

'There's no need to be like that,' the woman said, flustered.

'Mebbe not . . . but as far as I know you wanted nothing to do with the kid or her mother, and you did nothing for either of them. My missus and her friends have run up and down the country sorting things out. You've turned up too late, for me.'

'My wife's the next of kin,' the man said, squaring up.

'And my wife's the Queen of China. Now hop it. Like I said, get in touch after we've had a chance to talk, and we'll see what can be arranged.'

They went off down the path, muttering vaguely and turned at the gate. 'We'll be back,' the woman said. 'We know our rights.'

'God,' Kath said when the door was shut and the children had been diverted to the toy-box, 'what made you go for them like that?'

'They were treating her like a parcel,' Dave said. 'Walking in here and expecting to pick her up and take her off. I should co-co.'

'And Kath didn't like the look of them,' Jenny said breathlessly. 'She was upset on the phone.'

'Calm down,' Alan said. 'We have possession at the moment. Agreed, the situation is dicey: the authorities will have to be notified and some kind of temporary wardship arranged. But I can't see any court in the land awarding custody to unknown grandparents . . . not without some convincing proof of good intent.'

'Kath thinks they came sniffing round because they read all that rubbish about Elaine being Keir's lifelong friend.'

'If they swallow that they're probably too thick to have custody,' Euan said. It was an uncharacteristically waspish remark for Euan and Jenny realized how much Elaine's death had affected him.

'Well, if Alan's sure we're OK for the time being, I think we should all go up there as soon as we can and have a round-table conference.'

'I can make next weekend,' Alan said.

'And me,' Jenny offered.

'That's fixed, then,' Euan said.

'I'll drop Kath a line tomorrow and tell her what to do in the mean time . . .' He grinned. 'Although it sounds as though Dave is worth half a dozen legal levers.'

'Kath said L.J. just clung to him,' Jenny said.

'Oh dear,' Euan was looking rueful, 'I hope she's not getting too attached to him.'

'She probably is,' Alan said. 'But what's the alternative? Now, I don't want to rush you but the traffic's still quite dense and we did promise . . .'

'I'll get my coat,' Jenny said and made for the bedroom. Tonight they were to be front-row guests at Keir's first live show as producer/presenter. It wouldn't be fair to him to be late.

'Didn't Eleanor want to come?' Jenny said, as she sat between the men in the back of a black cab.

'She's in Winchester this weekend,' Alan replied. 'She has a sister there.' There was something in his voice that suggested he was not exactly pining for Eleanor but Jenny daren't probe. If it had been Euan she would have teased him about absence making the heart grow fonder, but with Alan you knew to keep your distance.

They gave their names at Reception and were ushered to a side door. A bright Toni look-alike appeared to take them to Hospitality. Perhaps every TV show had a Toni clone to do the rounding up.

And then Keir was swooping on them, greeting them with what appeared to be genuine delight. 'I'm *so* glad you three are here. I want you on the show soon, Alan . . . while I can still afford you. As for you, Jen, you must come over and work for me here.' He bent towards her and muttered, 'I wouldn't have to watch my back with you.'

They drank and talked, watching the famous and would-be famous throng the room and consume vast quantities of booze. Keir moved from group to group, sipping a scotch and water . . . 'The same one all the time,' Euan noted . . . and then they were being shepherded to their seats and the warm-up was beginning. 'Corny jokes and a little smut,' Jenny thought but it certainly got people laughing.

At last the floor manager was raising his hands higher and higher, soliciting applause, and Keir was striding down steps to stand in a spotlight that gleamed on his golden head.

'I'm glad you're all here,' he said, and then, leaning forward as though to be confidential. 'I'm not terribly sure of what we're doing tonight . . . they don't tell me much.

But you're here and I'm here, so we're bound to have a good time.' His smile was the smile of the little boy let loose in the cookie cupboard.

'It works,' Alan said quietly. 'You've got to admit he's good at it.'

30

3 March 1985

Kath and Jenny walked together between the neat lines of graves. Behind them Euan pushed Margaret's buggy and chattered to the two other children. Kath carried the container holding Elaine's ashes, but when they came to Jed's grave she handed it to Jenny.

'You do it, Jen. It should be you.'

There didn't seem any point in arguing so Jenny, who had carried it from London, took it back. Euan had parked the buggy and sent Brian and L.J. to run around the long circular path. Their laughter rang out in the quiet place, drifting back to the group around the graves.

'*John Edward Dawson, dearly beloved son* . . .' was written on the tombstone. There were primulas blooming on the grave, and two rose trees, neatly pruned.

'Should we say some words?' Euan asked. He had lifted his collar around his now-thin face, and his hands were thrust into his pockets.

'I don't think so,' Jenny said. 'We're here, that's what matters.' She let the dust drift out slowly, moving the box so that it fell along the length of the grave. 'I wish we could put her name there,' she said.

'You can't . . . because of Jed's parents.' Kath's tone was final. 'Too many questions,' she added.

The children had completed the circle and came to stand, breathless, by their elders, faces curious.

'What's this?' L.J. said, putting out a hand to touch the stone.

'It belongs to someone who went to Heaven,' Kath said. 'It's to remind us how much we liked them.'

'One person?' L.J.'s mouth was pursed for an answer.

'Two,' Jenny said. 'Two nice people.'

'Oh,' L.J. said. 'Good.'

Euan held out his hand and she took it, swinging on his arm and smiling up at him.

'Race you,' she said, and he took off, accommodating his long strides to hers, Brian bringing up the rear.

Kath watched them and then turned back to the grave. 'They're together now,' she said. 'Did she love him, d'you think?'

'I think she did in the end,' Jenny replied. 'I'm sure she did.'

They linked arms as they walked away, without turning to look back.

Steven had watched the scenes outside the TUC head-quarters as striking miners angrily urged members of the executive not to call off the strike. They were inside voting now, but the result of the vote was a foregone conclusion.

'They've got to end it,' Cissie said. 'Three-quarters of the men's back already. What's the point of dragging it out?'

It pained Steven to agree with her, so he stayed silent. The strike had been a mistake from the start – mountains of coal had been stockpiled, it was called at the wrong time, and a ballot, the miner's life-long privilege, had been denied him . . . if Scargill kept his King Arthur image after that lot, there was no justice.

Outside, the sun was shining. If the bairn had come round he might have taken her for a walk but there were London friends of her mam's down, so Cissie had said, and they wouldn't be seeing her all day. He got to his feet and went to the pantry, looking around in the packets and tins for a biscuit. There was nowt much else to do but eat nowadays. Sunday papers weren't worth reading any more: all tits and bums and criminals grinning out at you after selling their stories, as though they'd done something clever.

'You know something, Cissie . . .' She was chopping carrots and scooping the rings into a pan and once more he envied her always having something to do.

'What now?' she said. 'You've been sitting there like the crack of doom, so I know something's coming.'

'All right, all right, if you're going to be like that I won't speak.' She was on to turnip now, cutting wedges with a sharp knife. He listened to it thudding on the board and wondered what it would be like to chop.

'Come on . . . I'm listening,' she coaxed.

'It's nothing . . . it's just that there's nowt much to do on a Sunday now.'

'Oh, I wouldn't say that, Steven. There's the taties to peel, the puddings to mix, two jumpers to wash out by hand . . . you can have your pick of that lot.'

'Me do woman's work?' he said, taking his coat from behind the door. 'Hell'll freeze over before you catch me at that game.'

Euan and Jenny had travelled down together in Alan's car, and Jenny had invited both men to lunch with her parents. Alan had declined, pleading the need to sort out his house, but he would be coming over in the afternoon to meet Elaine's mother and stepfather.

'He'll fix them,' Euan promised Jenny. 'He can be really chilling when he likes. I've seen him in court several times now.'

At the lunch table Jenny watched Euan putting himself out to entertain her parents with tales of his stay in Zimbabwe. At times there were flashes of the old Euan, amusing and lightweight, but there was a greyness about him now, a fading of all that had been brightest.

'Poor Euan,' she thought, 'I hope he gets over Elaine eventually and finds himself someone nice. Someone who will take L.J.,' she added, as an afterthought. Kath would have to break the news of her pregnancy soon and that would surely mean L.J.'s days in Belgate were numbered?

'He's lovely,' her mother confided over the washing up. 'What does he work at now?'

'Nothing,' Jenny said. 'Well, he plays the drums but he's seriously rich. He came into a fortune.' She almost laughed out loud at the shine in her mother's eyes. 'Don't get carried away, mum. We're just good friends.'

330

'That's what the film stars say,' her mother answered. 'And then they get engaged.'

Kath had the dinner cleared by half-past one and checked the front room twice. Alan had asked for a place where he could be alone with Elaine's parents. 'And keep the children out of the way,' he'd said. Dave was all set to take them out before the confrontation began.

'I don't want them talking about Elaine being dead in front of her, Dave. I know she's got to be told soon, and I will tell her, but not with that lot here.'

'It'll be all right,' he said, as he went down the yard. His face above his parka was thin and worn, product of a year-long strike. Kath was suddenly conscience-stricken: she had been so taken up with her friends' troubles she had never given a thought to today's executive meeting and what it must mean to him.

'It'll be over today, pet. You'll see.'

'We may go back to the pit, Kath, but it won't be over. I've got a feeling it'll only just begin then.'

Alan and the others arrived five minutes later and she had tea already brewed.

'Only tea, I'm afraid.'

'Nothing nicer,' Euan said. 'But . . . I've got a bottle here for when we celebrate fixing the wicked stepmother.' The bottle was orange-capped and large.

'Very nice,' Alan said, 'but don't expect too much.' He turned to Jenny. 'I'm going to do my best, but the fact remains that she is L.J.'s only blood relative, and she has some rights. If she's the person you've described I'm going to try to persuade . . . or frighten her . . . into relinquishing those rights. If she can't be coaxed or frightened, we may have to go through the courts, which means one of us will have to apply for guardianship.'

Jenny saw Euan's mouth opening to offer himself but she heard only Alan's words: *'If I can frighten her . . .'*

'There's something I haven't told you,' she said. 'It just might help. Ages ago, before L.J. was born, Elaine told me why she left home. She didn't get on with her mother but there was something else: her stepfather sexually molested

331

her. She stuck it until she was old enough to go, and then she left.'

'Did her mother know about it?' Alan said. Kath's eyes had filled with tears and Euan's jaw had hardened so that he looked suddenly frightening.

'I don't know,' Jenny said. 'Elaine thought she probably knew but wouldn't acknowledge it.'

'That's quite common,' Alan said in response to Kath's gasp of disbelief.

'L.J. can't go to them now we know this,' Euan said. 'I'll take her to Europe . . .'

'Steady on.' Alan was smiling. 'Don't act as though I've lost the case already. I'm not bad at my job, you know.'

He spent twenty minutes in Kath's front room. Twice they heard the man's voice raised and the woman's voice grow shrill. They heard nothing from Alan until he led the way back into the kitchen.

'Mr and Mrs Foster have thought things over very carefully and have decided that they wish to renounce any claim on Lavinia Jane. I've explained to them that any compensation for the road accident will be minimal, since the driver had no insurance. They haven't the resources to give L.J. what they feel she needs by way of upbringing, so they have agreed to assign their rights to Mrs Gates in the first instance. I have suggested that they take legal advice on their own account, but they don't feel the need to do this. Now, if we could have their signatures witnessed?'

'He must have had this document with him all the time,' Jenny thought, looking down at the typed pages. 'Even before I told him about the child abuse, he thought he could win.' Alan was deep and ruthless, but that was what you needed in a tight spot.

As Kath ushered the subdued pair over her step Jenny reached to hug Alan and kiss his cheek. She felt his arm come gently round her and squeeze, and then let go. 'Thank you,' she said. 'For all of us, especially L.J.'

'Think nothing of it . . .' he said loftily and then, the old Alan, 'get the top off the frigging bottle, Euan . . . I think we can afford to celebrate now.'

'How did you do it?' Jenny asked as they drank.

'I told them Elaine had left a written statement about her stepfather's . . . activities. I told him I'd pursue him through the courts. He was as guilty as hell, you could see it on his face. And the mother said nothing.' So Alan had been happy to lie in what he saw as a good cause, Jenny thought.

She was sipping her way through the second glass when she realized they still had not solved the problem of a permanent haven for L.J. Perhaps they should have left the celebrating till later.

Keir poured another cup of coffee and padded back to bed. It had been almost dawn by the time he got home, and he had slept till long past noon. Since he woke he had drunk coffee and listened to the radio while reading and re-reading the reviews. All of them were good. 'Badly needed fresh talent' . . . 'his occasional gaffes only added to his charm' . . . 'we've waited a long time for a new TV personality: now we have one' . . . 'the women will swoon over him, the men will enjoy the wide sweep of his programme' . . . God, every time he read them he liked them better. The first show had even been reviewed in some posh weeklies. He had pitched both shows just right, and the wink at the end each time had been inspired. Even taxi-drivers were now doing it to him.

Money was now being waved at him from every quarter. He would soon need some professional help with finance. If he went independent eventually, the more of his own money he had, the better. But he wanted to hire Jenny: she had a good brain and she was loyal, the way she had stuck to Elaine proved that. Poor little Elaine . . . he must remember that child's birthday, or Christmas. Christmas was easier.

Anyway, if he got Jenny she would see to things like that. She could move in here – not into this flat . . . too confining – but into the block. He wasn't averse to taking up where they had left off long ago, but he would never make the mistake of committing himself again. Even if they only stayed occasionally, once they had their pills in your

333

bedside drawer and their Tampax in the bathroom cabinet you had hell's own job to get them out.

Keir lay back on the pillows, thinking suddenly about sex. These last few weeks it hadn't seemed to matter that much; it certainly hadn't been uppermost in his mind. There had been a kind of orgasm in hearing the applause, turning to see the floor manager's hands limp at his sides while the clapping swelled of its own accord.

The music on the radio gave way to a precise voice reciting the latest news. The miners' strike was over. God, there would be some beer drunk in Belgate tonight. He was slightly regretful that the miners hadn't won a famous victory, but they'd bloodied Maggie's nose just the same and driven Ian MacGregor to put his head in a carrier bag. Perhaps there was a programme in it? He reached for a pad and began to scribble.

'It's over, then,' Jenny's father said as she walked into the sitting-room.

'Yes, we got it all sorted . . . oh, you mean the strike. Yes, it's over, thank God.'

'They'll be singing and dancing in the streets through there, I suppose?' her mother said.

Jenny thought of Dave's face, almost sad as he heard the news. 'No,' she said, 'it's very quiet, actually. I suppose it hasn't sunk in.'

Up in her bedroom she drew the curtains on the darkness outside. If it had still been daylight she would have taken the dog up on to the hill and walked and walked. She wished Paul were nearer – perhaps they could both have sneaked out and met for a few moments? Except that he had a wife and children to keep him occupied: it was only she who had time on her hands in which to torment herself.

She tried to think about work. She had planned the programme around the Ethiopian famine, but she could rejig it to do a post mortem on the strike, instead. Industrial relations were Paul's sphere . . . but she had used him twice, and tongues would wag if she invited him back again too soon.

'We will always have to hide,' she thought. 'Not only to protect his marriage but to save his career.'

'*What about me?*' the voice in her head insisted, but she brushed it aside. She might take that job with Keir after all . . . TV would be a new challenge, something to take her mind off things. She rummaged in the wardrobe for her old, comfortable candlewick dressing-gown and wrapped herself in it. Tonight she would be the daughter of the house again and allow herself to be pampered. Tomorrow she would think about L.J., and Paul, and Keir's offer, and programme ideas for the rest of the run.

She caught sight of herself in the mirror, her brow furrowed, her mouth turned down. 'Life is a bugger,' she thought suddenly, remembering the fragile dust drifting down to spatter the primulas and settle on the quiet earth. Perhaps Elaine was the lucky one, after all . . . except that you had no choice but to go on and grasp the nettle of living. She brushed her hair back into place and went downstairs in search of some cherishing.

31

8 March 1985

'I ought to go,' Paul said, not moving in the warmth of the bed.

'So you should,' Jenny said, wriggling inside his arm. 'And you in the Whip's office, an' all.'

'Shucks,' he said and kissed her on the mouth. 'And now, Delilah, I *am* getting up.'

Jenny made wholemeal toast and squeezed oranges while the coffee percolated, and then sat down at the kitchen table to sift through her mail. Paul came, sleek from the shower, still buttoning his shirt with his hands, chewing on toast held in his mouth.

'Anything interesting?' he asked.

'Um . . . a letter from Barbara, my friend in Infants' School, remember? . . . and an invitation to her wedding,' she said, still reading Barbara's letter. 'She's having four bridesmaids, the gowns are modelled on Charles and Di's wedding, they're honeymooning in the Seychelles, and then . . . he's leaving Jersey and going into business over here! Back home, actually. I told you he was a solicitor, didn't I?'

'That's nice,' Paul said, wincing as too-hot coffee burned his tongue.

'No, it's not,' Jenny said. 'I think I've outgrown Barbara. She asks about our old friends, but only if she considers they merit her consideration. Keir Lockyer is in, and Alan, but after that – zilch.'

'I thought she and Kath were at college together?'

'They were. And she knew Elaine – although, to be honest, that pained her even then.' She picked up the deckle-edged invitation. 'July 27th: oh dear, I think I'm otherwise engaged.'

*

The invitation said 27 July. Alan put it on the mantelpiece for later perusal. He ought to go, for old times' sake, but he wasn't consumed with enthusiasm.

'I'm at Bow Street this morning,' Eleanor said as they went out to the car. 'And remember I'm away next week.' She waited while he unlocked the nearside door of the Rover and then folded herself gracefully into the passenger seat. That was what had first attracted him, the fluidity of her movements.

'Let's do something nice this weekend,' he said as he negotiated the narrow streets near their chambers.

'Like?'

'Oh, I don't know. Laze around, eat somewhere new, browse through a market, see a good film . . . they say *The Dresser* is excellent.'

'That sounds like a good plebeian mixture,' Eleanor said.

Alan turned to look at her profile. The mouth was smiling, so why did he feel put down? 'It was only a suggestion. You can choose. I thought that as I was away last weekend it might be nice to keep things easy this weekend.'

'When are you going to come home with me? You've had an invitation for a long time. I think they're beginning to wonder if your intentions are strictly honourable.'

'Are they?' he said laconically, steering the car into their private parking space.

'You should show your face, darling. And you'll like them.'

'Fix it up,' he said. 'Just tell me when,' and reached to pat her long, slender, ringless hand.

Kath looked at her list again. 'That's it, I think.' Mr Siddiqui went to the till and began to ring up her groceries. She was buying sparingly still, but at least she was buying with cash. And she was buying in the place where she'd met with kindness through a bloody awful year. A lot of people were going to the supermarket in an effort to spin out the first money they'd had in a long time, but she was not.

Mr Siddiqui handed her the goods and she packed them into her string bag.

'One more thing,' the shopkeeper said. He turned to the shelves and moved the tins, taking down a plastic bag folded into a small square.

'Remember this?' The half-sovereign was lying in his palm.

'I sold it to you.'

'Yes. And I am a business man, so when I sell it back to you I want my fair profit. On £60 . . . shall we say £2? Take it now, I'll be glad to be rid of the responsibility. Pay me when you're flush!' He grinned at the word. 'Flush,' he said again, and laughed.

Kath cried all the way home, bending her head to the pushchair as though to avoid the wind.

Dave was in the kitchen, drinking tea from a Sunderland Football mug. 'What's the matter?' he said. 'Who's upset you?'

'No one,' she said, brushing the tears away with her sleeve. 'I've just got back me faith in human nature . . . and while I'm on, Dave, you better sit down.' She saw his knuckles whiten on the mug.

'What is it?'

'Keep your hair on . . . I'm pregnant. Fourteen weeks. I wished I wasn't at first, but now I think I'm glad.'

Dave's head had dropped to his chest, as though she had delivered a *coup de grâce*.

'It's not that bad. It isn't due till September, and we'll be on our feet by then.' She waited, unable to bear his downbent head. 'Say something, Dave.'

What would she do if he said 'no way'? She couldn't bear to lose the baby . . . not now.

He looked up suddenly and she saw he was grinning. 'I can still make it, then?'

Kath felt a surge of relief that he was pleased. 'Who said it was yours?' she retorted, and let him haul her into his arms to be suitably chastized.

Euan was waiting at the table when she reached the restaurant. He had phoned her at ten o'clock: 'I've something to tell you, Jenny. Could we meet for lunch?'

Now Jenny studied him over the top of the huge menu.

He still looked washed out, but there was a more settled air about him, somehow. She quickly settled for grilled sole, then turned her attention fully to him. 'Now,' she said, 'what's this big news?'

Euan raised his glass.

'To us, Jenny. All of us, especially L.J.' They drank, and she waited quietly until he was ready to speak.

'I'm going to see Alan this afternoon; I rang him after I spoke to you. I want him to apply for custody of L.J. – on your part or Kath's, if one of you will agree; on mine if not. Although, as I'm a man, I may have less chance. Whoever does it, I intend to settle money on L.J., enough to see her well into her adult life. I've made enquiries, and there are boarding-schools that take children full time . . . mostly they're children of people living and working abroad, but they are good places . . . happy places. If we all rally round, and visit her and write to her and take her at Christmas and for holidays, L.J. can have a decent life. It'll be up to us to make sure she's happy.'

'Do you realize how much that will cost?' Jenny said.

'I can afford it. The money is the least important item, as far as I'm concerned. Once it's all signed and sealed, I'm going off to Eastern Europe. I can't stand London, now . . . not since Elaine died. And things are happening in Europe. This man Gorbachev, the guy who'll succeed Chernenko . . . I think he'll make a big difference.'

'If he ever becomes supremo,' Jenny said. 'You know the Soviets, if your face doesn't fit, you disappear. Besides, you're *not* thinking of going to Russia?'

He smiled at that. 'No, there isn't much call for rock drummers there. I'm going to Berlin, as soon as I can sort out what's happening to L.J. I can play in the clubs there and make a living, so they say. I know some guys who've done it. That will free my allowance for L.J.'

'Why go there now?' Jenny said, uneasily. 'It's too soon after Elaine. Wait a while, and see how you feel a little later.'

'I've *got* to go, Jen. You know me, strong as a chocolate soldier. I always run away.'

It wasn't true but Jenny could see Euan was not to be

dissuaded. So they talked of Europe while they ate, and then about the job Keir had offered Jenny.

'It's tempting,' she admitted.

'Will you take it, then?'

'I don't know. I'm going to sit in on his show tonight, to get the feel of the thing. I'll make my mind up after that.'

Kath had made her mind up at the tea-table, listening to L.J. prattle on about 'when mammy comes to get me'. Now that she had told Dave about the baby he was sure to press for L.J. to move on. It wasn't fair to ask him to shoulder another burden, not his own, and Euan couldn't foot the bill forever.

She couldn't bear to let L.J. learn about her mother from a stranger somewhere, and it wasn't fair to ask Jenny, who'd had no experience of kids. She must be the one to break the news. It was going to be hard enough to give L.J. up, but if she let the child go in ignorance she wouldn't be able to live with herself.

She put Margaret to bed early and sent Dave off to bath Brian. Usually L.J. and Brian were bathed together to save hot water but tonight she decreed separate baths. 'You're in work now. We can afford it.'

'Come here and give me a cuddle,' she said when they were alone, lifting L.J. on to her knee. She came willingly enough, clutching Brian's Action Man in one hand, his khaki combat jacket in the other. They played with the doll while Kath tried to find words that would not come. At last she sensed that L.J. wanted to get down and it stiffened her resolve.

'You know I love you . . . and I always will,' Kath said. L.J. nodded obediently and Kath bent to kiss the top of her head. 'But you know you only came here for a while?'

'Yes.' The child wriggled. 'Can I get down now?'

'Not yet, in a minute. I expect you'll be going away soon . . .'

'When mammy comes?'

It was now or never . . . and it had to be done.

'Mammy can't come back, L.J.' She felt the child tense but there were no tears, only two eyes fixed on her face and a deliberately expressionless face. 'Mammy wanted to come back because she loved you so much, but then there was an accident . . .' Her voice trailed away. How did you say 'dead'?

'Did she die?'

'Yes,' Kath said. 'She died, darling, and went to Heaven.' She hadn't meant to say that but it suddenly seemed essential.

'Can I get down now?' L.J. sounded so calm that Kath was almost shocked. She had expected every reaction but this, this almost unfeeling acceptance. It was then that she noticed the tremor in the child's body, the almost imperceptible rocking motion. She drew L.J. to her and kissed her again. The child didn't fully understand yet, but at least she had made a start.

'Yes, you can get down. And don't worry . . . when you go . . . well, when . . . we'll still keep in touch . . .'

'Kath!' Dave was there in the doorway, his towel-clad son in his arms, his face dark as a thunderclap.

'Me now,' L.J. said, lowering herself from Kath's knee.

Dave put Brian to the floor and held out his hand to L.J. 'You next . . . and I'm going to put soap up your nose!' L.J. squealed and ran past him.

'She stays here,' Dave said to Kath, and turned to follow L.J.

'Wait, Dave . . . do you realize what you're saying? We can't decide in one minute . . .'

'We didn't,' he said, starting up the stairs. 'I made my mind up weeks since.'

Jenny sat quietly behind the production team, seeing Keir's head larger than life on screen after screen. The production team were chattering away, each pursuing their particular function.

'We're running fifteen secs behind, Desmond.'

'Ready with the Astons, Carol?'

'Close in on three.'

'Sod, the flaming autocue's gone down. No, we've got it again. Stand by for VT.'

Keir was winding up an item on Beirut, sliding smoothly

from mayhem to music. A group appeared on the screen, four young men in black, stepping forward and back in time to their music. And Keir was speaking to the gallery on talk-back, checking on the next item.

'They've got this woman,' a girl said to Jenny. 'She's just given birth and swears she knew nothing about even being pregnant. We've got her down the line in Manchester, with husband and baby. Keir's got an obstetrician here in the studio . . . to comment. I can't say it's my idea of a hot item, but he's the boss.'

'He's a bloody law unto himself,' another girl muttered and then looked apprehensively at Jenny.

'Don't worry about me,' Jenny said. 'We're friends, but I'm not in his pocket.'

The music ceased and the mother came on to the screen, plump and smiling, a shawl-wrapped baby in her arms. She was pushing the shawl back from its face, turning it slightly towards camera as she spoke in her halting northern accent. 'So I never even thought it might be a baby, like. I just thought I were getting fatter . . . and he didn't help, allus saying things about eating too many butties. I went to the doctor, an' all, and he give me a diet sheet.' Beside her her lorry driver husband beamed proudly.

Keir talked to them earnestly, drawing out their total ignorance of what had been happening until the baby was almost in the world, making them sound unutterably stupid.

'They're a couple of morons,' the director muttered but he didn't sound too happy.

'Babes in the wood,' the producer said. 'I hope no one's going to blame me for using it.'

Jenny felt distaste rising in her throat. Keir, while posing as their understanding friend, was making them look ridiculous. And then he was turning from the screen picture of the parents to the obstetrician there with him in the studio, young and dark-haired with imposing, heavy-framed spectacles.

'Is this possible, doctor? It sounds almost unbelievable.'

The specialist smiled thinly and Jenny felt her skin prickle. 'Possible, perhaps. Probable, no. We usually find in these cases that the mother has deliberately blocked out the fact of her pregnancy . . . seeing what she wants to see rather than what is. That happens more often than you might think.'

'But why would she do that?'

'Because she doesn't want the baby. For some reason of her own, perhaps a compelling reason in her view, she sets her mind against the growing foetus . . .'

Suddenly the mother's face was on the screen, her head shaking from side to side in denial, her mouth trembling as she clutched the baby she had so proudly displayed a moment before.

'Get off her!' the producer screamed as the distraught mother turned to the father, her lips moving in what was clearly a plea for him to put the record straight.

'What do you want?' someone was shouting. 'Where the hell do I go?'

'Get off her . . . get *off* her . . . give me the fucking studio floor, but get off her face!'

And then Keir was there again, blond and smiling. 'So our "unexpected event" probably wasn't as unexpected as we've been led to believe . . .'

'Christ!' someone said.

Jenny got up and moved backwards, reaching for bag and coat as she went. She let herself out of the gallery and then started running for the foyer and some cold, clean air.

Beside him Cissie had stopped sucking her boiled sweet and was leaning forward, face intent. Steven stood up and made for the door.

'I'll put the kettle on,' he said by way of excuse. 'He'll be done in a moment.'

It the kitchen he filled the kettle and set it on the flame, trying not to think of the woman's tearful face, of the babby on her knee, the husband shifting miserably beside her. Why did people ever agree to go on television? It was their own fault if they said yes.

But his own face looked back at him from the uncurtained window, a ghastly incomplete image but one that could not be ducked. The kettle gathered steam, beginning to bounce gently on the stove, and from the room behind him he heard the Keir Lockyer Show's signature tune, marking the programme's end. Keir would be standing there while all the names rolled up, closing his eye in a cheeky wink and turning his head as he did so by way of a goodbye. Making on he was still a canny lad.

As the kettle began to whistle Steven leaned forward, smiling wryly, and addressed his reflection.

'Steve, lad, thy son's a shit!'

BOOK FOUR
1989

32

19 January 1989

'I think that's all,' Jenny said to her secretary. 'See if you can get Ludovic Kennedy and Robin Day . . . there's this interesting thing between them over the biographies they're writing. I'll speak to them, if there's any chance they'll do it. I'm going to have coffee now and then talk to Chris about next year, so no calls unless it's a fire. Oh, and try and get Simon Weston for the 2nd. He's so good.'

She took coffee from the percolator in the corner and went back to her desk. She hadn't had time to read her mail this morning, she had simply tipped it into her huge leather bag. She sorted through it now, dumping the junk mail unopened, putting aside a letter addressed to Paul, opening an airmail envelope addressed in Euan's flamboyant hand. She had always enjoyed hearing from him, and with so much going on in Eastern Europe his letters were now of added interest. He was in Prague at the moment, and she read eagerly on past the usual good wishes and personal details.

I walk amid squadrons of old stone saints, trying not to see or hear the police cars belching out orders. The villages outside are from another age, folk dances and craftsmen and all so Catholic. If they're Slovak, that is. I'm still struggling to understand the differences of race. We're such a small breed, Jenny, we humans . . . why do there have to be so many divisions amongst us?

Everywhere you go you hear Gorbachev, Gorbachev; his name seems to be on everyone's lips. He can't possibly fulfil all their expectations, Jen. I felt hopeful at first, but now I think they're asking too much, too soon. The police are everywhere, in spite of Czechoslovakia signing the Helsinki Agreement. Human rights don't seem to matter here, agreement or not. Charter 77 tries its best, but they are arrested if they say too much, God help them. Still rock and roll is big here, so I will stay for a while . . .

He ended with the usual promise to come home soon, but Jenny didn't expect him. Perhaps he was a latter-day Flying Dutchman, doomed to sail the world forever and never find a home. He had settled money on all Kath's children when she and Dave had finally adopted L.J. And by now he would know Kath's news, although he hadn't heard it when he wrote the letter. She put it back in its envelope, a little knot of pain inside her at the thought of Kath. She must get up to Durham as soon as she could, before it was too late.

She opened the *Independent* while she drank her coffee, flipping through Home and Foreign news, knowing she must digest it later, and looked at the Gazette section.

It almost escaped her notice. *'Wentworth. To Guy and Barbara (née Finch) the gift of a son, Sebastian Miles. A brother for Charlotte Anne.'* So Barbara had had a second child. Kath now had four and Barbara two. Elaine had had L.J. Only she had failed to reproduce. She badly wanted a child . . . Paul's child. They had talked about it more than once, but the risks were too great. When his children were grown and he could decently leave, she herself would be forty. Too old to start a family?

Jenny folded the paper and stood up, knowing where this train of thought would take her. She had work to do, important work. 'I'm a bigwig,' she told herself, laughing. When she ceased to find her exalted status funny she would get out. If you began to believe your own myth . . . She picked up her notes on the 1990–91 funding and left her office.

'Of course you'll miss them, Steve. I'll miss them, never mind you. But life goes on. There's plenty bairns around.'
'Not like that bairn.'

348

'No, I'll give you that,' Cissie said. 'She's got a bit about her . . . and bright! I'd like to be a block behind her.'

She passed him a cup of tea and hitched her chair closer. 'I've been thinking, Steven. When they go, we could get a dog. Not a pup . . . so don't say you won't live long enough. But they have lovely dogs down the pound, grown-up dogs. House-trained.'

'I don't care if they're circus-trained, Cissie, they are not coming in here. Not in my half, anyway.'

'That's it, be spiteful! You know I couldn't manage one on me own. You're turning spiteful, Steven Lockyer. Our Keir says you've changed . . .'

'How would he know?' Steven said. 'We only see him when there's an X in the month. It's not me that's changed, Cissie, it's your precious son.'

'There you go again – "your" son! Always mine if he's not keeping up to scratch. Who was his father? The milkman?'

'For all I know.' Steven knew he'd gone too far before the words were out, but he could hardly take them back. 'Can it, can it,' he said over her protestations. 'I'm off down the club – the only place . . . the *only* place I get any peace.'

'That's right, get down there with all the other Buddhas. Sit and put the world to rights and get cirrhosis of the liver – see if I care. And when you're gone, Steven Lockyer, I'll fill this house with dogs. Fill it!'

He turned at the yard door, unable to resist it. 'You'll need a big shovel for the shit, Cissie,' he said, and ducked as the cracket cushion sailed past his head.

Keir changed jackets three times before he was satisfied. Things like that did matter. He wanted this new job . . . needed it even. You could live on doing voice-overs for advertisements but there was frigging little satisfaction in them. If Kepier hired him, he would try for a clothes allowance and other perks rather than a big fee. It was when the salary figures got too high that they started thinking. You could price yourself out of the market . . .

and perhaps that was what he'd done with Global? Or did he detect Annie Love's finger in that particular pie? In the beginning his show had been a hit – a mega-hit. He had stood in a W. H. Smith once and seen his face on the covers of three magazines, three at the same time! And then it had all started to crumble. People left the show . . . talented people . . . people he'd trusted. It had been harder and harder to keep continuity. Guests had become hard to persuade on to the programme, too . . . guests of consequence, anyway. In the end he'd been interviewing anything they could scrape up, and not even he could lift an entire show. 'I was let down,' he thought bitterly. And it *could* have been down to Annie Love . . . she had still been in Africa during the first series, but by the end of the second she was in Britain again, worming her way back into Light Entertainment. 'I should have crushed her altogether,' he thought. 'I didn't finish the job.' He had been without a slot of his own for a year now, and even one-off shows were fewer and further between.

He made the cab circle the block before he went into the restaurant. Mustn't appear hungry: if they thought they had you by the balls they lost interest. He would use the line about the book – he did have a publisher lined up, so it wasn't a lie. Something might even come of it eventually. These things took time.

He stood in the doorway, scanning the tables, hoping to see Kepier's sleek head. He wasn't there, and disappointment filled Keir. He would be there first and have to sit waiting as though he had nothing else to do.

'Can I help you?' The head waiter was staring at him. 'It's Mr . . .?' If he couldn't remember someone who had almost lived in his bloody restaurant . . . 'Why, it's Mr Lockyer, isn't it? Forgive me, sir, the name escaped me for a moment but of course, the face . . .'

'I'm meeting Simon Kepier.'

'Ah, yes . . . the corner table.'

Keir ordered a spritzer and drank it slowly. It was ten past one already. Perhaps Kepier wasn't coming? No, mustn't be paranoid. Traffic was the pits around lunchtime. He saw Kepier appear in the doorway and half-rose.

'Keir . . . God, I'm so sorry! Have you got a drink? Now

350

let's order quickly. If you don't mind I won't have a first course, I am *so* pushed for time. But you have whatever you want. You won't mind if I have to dash off quite smartly . . . I'd have rung you, but it's ages since we met and I thought old Keir will understand . . .'

In the old days Keir would have pleaded a forgotten engagement, and left Kepier to it. There was nothing to do now but smile agreeably, order what least repelled him, and move it from side to side of his plate.

Most of their furniture would be going in containers, but she needed to pack carefully for the voyage, Kath told herself. They were leaving on 22 February, five weeks from now . . . plenty of time if she thought it all through.

There were times when she couldn't believe she'd agreed to go. Gradually, ever since the strike, bitterness and disillusion had grown in Dave as pits closed and the union crumbled. Every time Arthur Scargill strutted the stage Kath saw her husband's disaffection deepen. A third term of Thatcher had been the final straw.

He had broached it one night: 'What about Australia, Kath? It would be a good life for the kids.' Kath had agreed to think about it, sure that he would drop the idea, given time. Brian and L.J. would be moving to the comprehensive in little over a year; Margaret was settled in the Juniors and Christopher would start nursery school at Easter. She couldn't uproot them and take them to the other side of the world, where she wouldn't know a soul!

But she was. Well, at least she would have her bairns, and Jenny had promised annual visits. If she'd been close to her mother, Kath might have been able to resist Dave's urging, but they had never regained their old loving relationship. 'We're polite with each other,' Kath thought, 'a bit like colleagues not wanting to tread on one another's toes.' Could she ever shut her heart to her daughters, no matter what they did?

She and Dave were happy, that was one thing. And they'd done well, really – the house would fetch at least £21,000, four times what they'd paid for it, and Dave had been snapped up by the Aussies as a skilled man.

She heard an altercation in the back yard and moved to the window. Christopher and little Joel from next door were pulling a transformer apart by traction, each wanting to possess it.

'That's enough!' she shouted, rapping on the pane. Her younger son turned and smiled at her. He would break some girl's heart one day with a smile like that, little bugger.

She turned back into her kitchen and switched on the radio. 'Riot police have occupied the centre of Prague, using tear gas, water cannon and baton charges to disperse demonstrators, mostly students, chanting the name of Mikhail Gorbachev, the Soviet leader. The demonstrators have taken to the streets to commemorate the death of Jan Palach, the student who set himself alight twenty years ago in protest against Soviet invasion. A Foreign Office source said . . .'

She looked out of the window again as a child began to cry, hoping that Euan would have the sense to stay out of trouble, for L.J.'s sake if not his own. His last card was on the mantelpiece, from whence it would join the collection in the half of the bedroom L.J. shared with Margaret.

'For God's sake watch it, Euan!' she said aloud, and went outside to declare an armistice.

Jenny, listening to the news, also thought of Euan on the Prague streets. Silly to worry, though – he had always been a roamer, and he always survived. She checked the pie in the oven, browning and rising nicely – Paul loved steak and kidney pie – and went back into the living-room, carrying salt and pepper, just in time to hear news of the death of the Shadow Minister for the Environment. He had only been forty-seven and had suffered a cerebral thrombosis. Politics was a lethal business. Paul was only thirty-nine, but at times he looked older. Her own job was stressful but not to be compared with his.

Now that Jenny had finally abandoned the idea of working in television, she was beginning to enjoy her work. Radio was a medium that had never been fully exploited. Perhaps she might push out a boundary before she gravitated upstairs and retired with a pension. But she must be careful not to lose touch with reality: if you worked in the media it was all too easy to see your job as the hub of the universe, to imagine that everyone in the world was holding their breath for an invitation on to your programme. But the world was growing used to TV and radio; people were becoming more sophisticated. 'One day they'll tell us what to do with our programmes,' Jenny thought and grinned at the prospect.

'Did you hear about Graham?' Paul asked when he got in.

'Yes. I'm sorry. I only met him once or twice but I liked him. Who'll get his job, do you think?'

'Oh Jenny, Jenny, I thought politicians were body-snatchers, but media folk have them beaten! Let the guy get cold before you fill his shoes.'

'You're right. But it's living with you five days a week that's changed me.' She could smell the pie in the kitchen and she knew she ought to see to it, but Paul was putting down his glass and reaching for her.

'It's Thursday,' he said, nuzzling her neck and then the cleft between her breasts.

'So?' she said.

'Friday tomorrow.' She knew what he was getting at but it was fun to play dumb.

'And Saturday after that. Big deal!'

'What does Friday mean?'

'Laundry night?'

'No. Try again.'

'Wogan on telly?'

'Wogan's always on telly. Friday night means we won't be together for three nights.' His hand was cupping her breast then moving down to pull her shirt from the waistband of her skirt.

'Paul, the pie!'

'To hell with the pie.'

'It's steak and kidney.'

'Who wants steak when he can have caviar?'

'Jeez . . . and you a good socialist!'

Paul let her go to turn out the cooker, then, and put out the lights under the vegetables.

They made love, slowly, the way lovers of long standing do. Why had she ever thought frenzy with Bart desirable? What could beat this slow, thrusting pleasure, his body moving above her and then, with a twist of his arms, moving below?

'What are we going to do, Jenny?' he asked when it was over. She knew what he meant but she couldn't discuss the future now – not without crying.

She stood up from the bed and reached to haul him to his feet. They stood naked, touching and fitting everywhere, her head beneath his chin, her arms meeting around his waist.

'I love you, Jenny.'

'Love me, love my pie,' she said. 'Let's go and eat.'

33

2 February 1989

The call had come out of the blue. 'Keir Lockyer? I'm Judith Gilham. I'm with Windswept . . .'

Keir knew Windswept, one of the better independent television companies. His heart began to thump uncomfortably but his voice stayed laid back.

'Oh yes. What can I do for you?'

'I'm compiling a series of profiles, and we'd very much like you to be one of the subjects . . . if you're free, that is?'

When he came off the phone Keir had had to resist the temptation to fall on his knees. This was it, the mega-blast of exposure that would put everything right.

Now he looked around the flat, checking that everything was OK for her visit. The stark black-and-white decor that had been so stunning when it was done was a bit *passé* now, four years on. He had added a few touches of turquoise and grey to take the effect up, but the new artefacts had highlighted the slight shabbiness of the rest. As soon as he could afford it he would gut the place and have it redone.

Money was running out faster than he would have believed possible. Still, he mustn't think of things like that now. She would expect him to fizz and he mustn't disappoint her.

Judith Gilham was rather a disappointment to him, when she turned up – short and dumpy, with unkempt mousy hair and the statutory uniform of media women: dusky black and faded denim. Curled up on his sofa, though, clutching a cup of coffee that she swore would 'save her life', she was more endearing. She wanted to know all about him, everything . . . from the beginning.

'I was born in County Durham, in Belgate. It's a pit

355

village. Dad was a miner, very socialist – still is. Mention Maggie Thatcher and his ears steam. We didn't have much . . . mam made her own bread . . .' Keir had a sudden uncomfortable vision of the Hovis adverts, a kid toiling up a bank against a sepia background. Mustn't be too plebby. 'When I went to Oxford . . .'

'I don't think I've talked to Keir more than five times since we came down from Oxford,' Alan said. 'And lately he's seemed to slip out of the limelight, so I don't see him on TV either.'

'I'm thick with his mother, Cissie,' Kath said. 'She's been a godsend with the kids and everything. She talks about him as thought he was, well, a good loving son, but I think they hardly ever see him. He doesn't seem to care about his roots much.'

'I think he never did, really.' Alan's brow was wrinkled in an effort to be fair. 'When we used to talk, all of us, spout politics and everything, he came out as the red-hot socialist – mining stock, up the workers . . . In reality I suspect it meant very little to him. He was simply marking time until he could get away.'

'And now he's a professional northerner,' Kath said. 'With that dreadful wink . . . I'm one of you, it's supposed to mean. Jesus!' She stood up. 'More tea?'

'Ta,' Alan said. 'I'm glad you were in. I couldn't come up here and not see you, not once Jenny told me about Australia.'

'You still come back here, Alan . . . you've kept on the house. Does it mean a lot to you?'

'Yes. It's strange – in the old days I could hardly wait to get away, I didn't feel I could function up here, I craved London and everything it represented. Now, I think of the old house as a kind of eyrie, somewhere I can fly back to whenever I get the chance. I think my mother intended me to sell it and use the money for my future.' He paused for a moment. 'I think that's why she did what she did.'

'She loved the house but she loved you more?' Kath said.

'Yes.' He smiled. 'She thought I belonged in London, but really this is where I belong.'

356

'Jenny's like that. I think she truly loves Durham and Sunderland. In her letters sometimes . . . I know we've got the phone but letters are so satisfying . . . in her letters, when she gets on about this area she's quite lyrical. Elaine used to say Jenny should write a book one day.'

'Perhaps she will,' Alan said and Kath threw back her head to laugh.

'You have changed, Alan, although you may not realize it. "Perhaps she will": in the old days you'd have said, "Maybe she will". True?'

He considered. 'Yes, you're right. Blame it on my legal training. Pedantry is an official part of the syllabus.'

He asked about L.J. and the other children before returning to the old topic. 'What about you, Kath? What about your roots when you up stakes and go to Australia?'

'I've thought about it a lot – what I'll lose, what I'll gain. But I really made my decision a long time ago. My place is where Dave is, or where he wants to be. It's corny, and it's not what I used to think, but it's right for me.'

'You *were* pretty militant in the old days. Well, feminist. Much more than Jenny – and yet she's turned out to be so independent, so strong.'

'She *is* a modern woman, Alan. Not strident, but she knows her worth. I'd envy her if the kids ever gave me the time.'

As Alan was leaving he kissed Kath on the forehead and then the cheek. 'Take care of yourself . . . and the kids. Give Dave my regards and say I'm sorry I missed him. I'll come back before you go, if I possibly can. If not, don't lose touch.'

He paused on the step. 'Do you remember what Jed said that New Year?'

'He said, "We are the future,"' Kath answered, instinctively knowing what was in his mind.

'We are the *bloody* future,' Alan corrected. 'What would he make of us now?'

Steven had walked Christopher to the school gates to meet the others coming out. The little boy shrieked with delight as his sisters appeared to make a fuss of him. Brian condescended a nod but was too lordly to walk home with them, swooping off with his friends.

'You don't smoke, Uncle Steve, do you?' L.J.'s eyes were round in anticipation of the wrong reply.

'I used to like a pipe of baccy. I don't now.'

'Good. Because it's very bad for you. We made a big chart today to show what it does, and it slows your legs down. You couldn't walk our Christopher if you smoked.'

'Yes, he could,' Margaret said. 'If mam said he could, he could.'

'Shut up, Margaret, you don't understand,' L.J. said kindly. 'I'll explain after.'

'OK,' the younger girl said. 'But I still think he could if he wants.'

Steven smiled to show he was glad of her back-up, but he felt far from cheerful. What would he do when they were gone? Cissie talked of making a trip to Australia, and Kath had said they were always welcome. 'But I'm far ower old to go round the world,' he thought. He could hardly make it as far as Sunderland now when Cissie wanted a shopping trip, never mind the bloody Antipodes.

Kath was in the kitchen when he got home, eating Cissie's bread buns.

'Your friend got away, then?' he said, while she fussed over her children.

'Alan? Yes. It was lovely talking to him again, though. Thanks for the peace and quiet. He was at Oxford with your Keir, you know – they used to come up and down together. He still sees him sometimes . . . now and then.'

'He's lucky,' Steven said. Moving to his chair he picked up the paper to show he had had enough of harking back. He was shaking it out to find the racing page when he heard Cissie.

'If only our Keir had wed you, Kath. If only he'd had the sense!'

'Oh, he was too go-ahead for me, Cissie. I could never have kept up with him.' Kath meant to sound humble but Steve could hear it in her voice, however much she might try to hide it . . . distaste for Keir Lockyer and all he stood for.

*

358

Jenny sipped a spritzer and tried not to fret. This was what Paul's life was like, lived constantly on a knife edge. The waiter was hovering, wondering if she wanted to order. 'I'm afraid Mr Kovalski's held up, Jean.'

He nodded sympathetically. 'I understand, it is politics. So many of our patrons are in this . . .' He chuckled. '. . . this line of business. We can always accommodate them.' He smiled and moved off, leaving Jenny in peace to reflect that he obviously accepted her affair with Paul. He had talked to her as though she was a wife. Perhaps she was – the weekday wife, just as Rose was Paul's consort at weekends.

'I have never had a Sunday with him,' Jenny thought. 'Not a whole Sunday, not a long, slow Sunday morning with nothing to do but read the papers.' Once they had planned a weekend, but his son had developed mumps and a row had blown up in the constituency over local council business. 'I *can't* leave Rose to cope alone,' Paul had said apologetically and Jenny had agreed.

She minded about weekends but there were other things she minded more. Whenever they went to functions they had to arrive and leave separately in case of cameras. At first it had been exciting to have him slip out of the taxi around the corner while she swept on to the venue alone. Now it was tedious. It was the same at the end of the evening – one or other of them took their leave and waited a hundred yards away. Farcical and increasingly degrading! She thought of him in her bed in the mornings . . . their time for making love because he got back from the House too late and too tired. Could she go on like this? Could she live without him? There would be other men, but would they do?

She was still seeking an answer when Jean reappeared, carrying a portable phone. 'Thank you,' Jenny said and waited to hear Paul's reasons for his non-arrival.

'I can't get away now, Jenny. There's been a reshuffle, and Bob Cardus is going to replace Graham. There's going to be a general move up, and Neil wants me to go to Environment. I love you. I'll see you when I can.'

In the cab Jenny realized that she had found her answer.

Paul was moving further towards centre stage, and there could be no room for her in the spotlight.

Kath tucked in a stray foot, an uncovered shoulder, and kissed the brows of all her sleeping children. Only L.J. was awake, bright-eyed and talkative.

'I'm glad you and dad don't smoke.'

'So am I, pet. I couldn't afford them.'

'No, I'm serious.'

'So am I. Now go to sleep.' She was at the door when L.J. called her back.

'Do you know why I'm glad?'

'No.'

'Because I love you. Both of you.'

Kath was half-way down the stairs, trying to sniff back tears, when Dave appeared at the foot, his face aghast.

'It's your mother! She's in the front room.'

'Oh God, what's the matter . . . it's not me dad?' Only death would bring her mother here after all the years of trying to pretend this house did not exist. Kath clutched Dave's arm when she reached him, and hissed, 'Make some tea.'

'No fear, Kath, I'm off down the club. I had enough of the battle-zone in the strike. This is strictly your war. Different if it was your dad or the lads, but not your mother – I know when I'm beat.'

'Hello,' Kath said, trying not to sound nervous as she came face to face with her mother.

'You're definitely going then?' The older woman was standing with her back to the fire, arms at her sides.

'You know we are. I told you ages ago.' Surely her mother wasn't going to cause a row now? She'd probably be glad to see the back of them. 'I wouldn't think you'd be bothered, any road,' she thought, but aloud she said, 'Want some tea?'

She brewed tea in stony silence except for an odd remark about the weather or the state of the road outside where the Gas Board had been working. She used cups and saucers instead of mugs, feeling that this was a momentous occasion but not knowing why, and put chocolate bourbons and gingernuts on a plate.

'Ta,' her mother said and sugared her tea. She looked around her. 'You won't be able to take all your stuff.'

Kath bristled. Was the purpose of the visit to scrounge? Surely not. 'You can get a lot in containers,' she said shortly and hid her disapproving mouth behind her cup.

'I'm glad you're going,' her mother said abruptly. And then her voice broke. 'Thank God you're getting out of Belgate.'

'You mean you're glad to be getting rid of me?' Kath said, not wanting to sound bitter but unable to adjust to this change in her mother's attitude.

'No, it's not that!' Her mother's face was anxious now, her fingers nervously twisting together. 'Can't you see . . . don't you understand? . . . I hated you for throwing away your chances, but it was because I loved you best of all.'

Kath shook her head, uncomprehending.

'I love the lads,' her mother said, her fingers fiddling with the buttons of her coat. 'You know I love them. But they belong here, like your dad. They're happy; they always will be. But you're like me, Kath. You can better yourself . . . like I could've done if I hadn't fallen with a bairn. That's what chained me to Belgate. I wanted better for you.' Her head dropped. 'That's all it was,' she said, quietly, the passion gone out of her. 'I wanted to give you the moon, and when I couldn't I turned against you.'

Kath hesitated, anger still in her, but only for a moment. 'Oh mam,' she said at last and saw her mother's head come up as she stretched out her arms. Kath went to her then, still not understanding fully but sure that her mother's arms were the best place in the world to be. She brushed against the table as she moved and heard a cup rattle in a saucer but it seemed not to matter. And then they were together, both of them half-crying and telling one another not to be so foolish.

'You're going to a good place, Kath. A new world.' Her mother's hand was soothing her brow, the gesture she had longed for so many times in the last few years. She felt tears come and then sobs were racking her.

'There, there,' her mother said. 'There, there. It'll be all right now.'

34

22 February 1989

They walked from room to room, making sure nothing vital had been left behind.

'I can't believe the time's come, Jen,' Kath said. In a little while they would set off on the journey to London, and then the long flight to Australia.

'You'll be all right,' Jenny said soothingly. 'You could cope anywhere, Kath.' She pulled a face. 'I wish I had the knack.'

'There's nothing wrong with you that a little effort won't fix,' Kath said. 'You've come home to lick your wounds and replan, haven't you? When you go back you'll lay all before you . . .' Realization jolted her painfully. 'And I won't be here to see it. My God, Jen, if you get married I won't be there!'

'That's the least of your troubles, Kath. I don't think I'll ever get married. I haven't been notably successful up to now.'

They had paused on the landing, looking from the uncurtained window to the pit stock towering above the roofs of the houses.

'You were keen on Keir once, I thought. Weren't you?'

'For a bit. I grew out of it.'

'Good job. I was all ready to land you one if you didn't. I had Master Keir sussed even then.'

'He isn't that bad, Kath. Just not what I used to think he was.'

'You know something,' Kath said as they moved down the stairs. 'Of all the four boys, the one with the most potential was Alan.'

'Yes, he's done well,' Jenny said.

'I didn't mean professionally,' Kath said. 'I meant

362

between the sheets. Mark my words, for all Keir's strutting about, Alan's the red-hot lover. It's in the eyes.' She turned and looked at Jenny on the step above her. 'Honestly Jen, if you hadn't had a lover for the last five years I'd think you were *virgo intacta*. You've gone pink at the very idea of Alan having carnal capabilities.

Jenny pushed Kath forward. 'Wash your mouth out, Kath Gates. If Alan could hear you now he'd have you for tort or something equally nasty.'

'Yes please,' Kath said. 'And now will you tell me why we're wandering around talking nonsense when I've got to up sticks and cross the world in five minutes' time?'

'I might be leaving England myself if I take that American job,' Jenny said.

Two newly delivered letters were lying on the mat, one a brown envelope marked British Gas, the other bearing a foreign stamp.

'Euan,' Jenny said. 'He never forgets, does he?'

Kath sighed. 'And neither does the Gas Board.'

'And then, of course, I worked with Annie Love.'

The researcher looked up, chewing her pencil. 'How did you get on with her? They say she can be difficult.'

'Oh God, I worshipped her. Of course she could be difficult, name me a creative person who isn't difficult. I can be a shit, I'll tell you that *on* the record. No, Annie was tops for me. When they told me she was going on to something else I begged them. "Don't do it," I said. "Leave Annie and me together . . ."'

She had ceased to write and was listening politely. Time to move on. 'Then of course I had the whole responsibility. Before that I'd been the front line, but now I was vanguard, rearguard, sometimes I even had to be the bloody paddy-wagon . . . But that's my style – get in there and be part of it. The Keir Lockyer Show was essentially a team effort, I can't stress that too strongly. When we get my new project off the ground that will be a joint operation, too.'

'What *is* the new project?'

He smiled ruefully. 'I wish I could tell you. I'm so enthusiastic about it, I'm dying to tell you, believe me. But the money men say it has to be kept under wraps for a while longer. All I can tell you is that it'll be *big*.'

They had talked for five days now. Surely she would tell him something soon, talk about dates? Even then he couldn't be sure the programme would actually be made, but it would be progress to know they were talking logistics. He needed to be before a camera again, so that he could talk to his heart's content without having to defer constantly to so-called experts. He was the expert, in this case; the only man who knew what was in Keir Lockyer's heart.

'I'll write – and I'll phone,' Kath said, holding her mother close. 'And you'll come out, you and dad? For a wonderful holiday, maybe even to stay.'

'We'll see,' her mother said. 'We'll see.' Her father was wiping his eyes and Dave's mother was crying openly.

'Time to go, pet,' Dave said in her ear. 'We can't draw it out, for their sakes.'

'I know,' she said, wishing with all her heart that she could wake up and find herself dozing by her own hearth and Dave coming in at the door, ready to eat his meal.

'Goodbye, Jen, God bless. You'll come and see us soon, won't you?'

'Next year,' Jenny said. 'That's a promise. Maybe sooner. Take care of the kids, and don't let them forget . . . not us, or Durham, or any of it.'

'Fat chance,' Dave said, taking his weeping wife's arm.

'God bless you, Kath,' Cissie called out.

'Bye-bye, Auntie Jenny,' L.J. said, waving through the back window. She had her pearls around her neck, the pearls that had been Elaine's. Flaky now, but still lustrous.

'God bless, L.J. You write!'

And then they were gone and the cries of 'phone' and 'letters' and 'holidays' died away, leaving those left behind to disperse to their homes to finish their weeping.

'I will not cry,' Jenny told herself as she climbed into her car but she did cry, half for the loss of Kath, half for the emptiness of her own life. She drove into Sunderland and parked in the town centre. If she went home now and her mother as much as put out a hand to console her, she would disintegrate. Better to stay among the crowds for a while.

She moved from shop to shop in the new Bridges shopping centre, buying a pink leather handbag and a pair of ear-rings, nostalgic for the days when she had been a student and couldn't afford to buy a paper bag. Eventually she found herself in Dunn's, the old-established department store. It was half-past eleven, and she could do with a coffee.

She had collected her cup and was looking for a vacant table when she saw Barbara seated in a corner, surrounded by other women, all of them dressed in expensive suits and white shirts, their faces porcelain, their hair like glossy helmets, their ears and fingers weighed down with jewellery that had the gleam of the real thing. They seemed to be talking animatedly, their laughter tinkling over the teacups. 'Jesus,' Jenny thought, 'they look exactly like the Stepford wives.'

She was turning on her heel to escape when Barbara spotted her.

'*Jenny!*' She was jumping up from the table and holding out her arms. 'Jenny, how fantastic!' She turned back to her friends. 'This is my very best and oldest friend, Jenny Sissons, who is very high up at the BBC and so brainy I can hardly keep up with her.' She looked around. 'There's a table, Jenny. Let's grab it and exchange our news. Back in a moment,' she trilled to the others, and led Jenny to a seat.

'Now, let me get my news out of the way and then I can hear all about you. I've got you on my list for a letter, but meeting face to face is so much better.'

She was blooming, the children were thriving and Guy was powering his way to the upper echelons of the legal profession. 'I've been so lucky, Jen. I've got everything I ever wanted. Well, almost everything. You always want just that little bit more, don't you?' There was something wistful in her voice, a note Jenny could not remember hearing from Barbara before.

'So you're happy?' she asked.

'Of course I am,' Barbara said vehemently. Too vehemently, too quickly, Jenny thought. 'I'm a *very* lucky woman, I've got everything,' she said again. Her pink-tipped fingers ceased to move together and balled into fists. 'Now,' she said brightly, 'what about you?'

'Kath left today. For Australia,' Jenny added, seeing Barbara's blank expression.

'*Australia* . . . poor thing! Still, he'd never have made his way here, so I suppose it's for the best. You've done well, Jenny, I'm always boasting about you. And Alan's getting to be quite well thought of. Guy's brother is a Recorder now and he says Alan is destined for the top. He's split up with that Eleanor, did you know? I met them in Harrods once. And what do you make of Keir? I used to like his programme but Guy said it was trash, so we didn't watch it any more. I never missed it,' she added. Again that almost wistful note in her voice.

'You knew Kath and Dave adopted Elaine's daughter?'

'Did they? Well Kath always was a bit . . . I was going to say feckless but of course it's a *good* thing they've done. Poor Elaine, I liked her so much, but she was doomed, wasn't she, once Jed went? And the child . . . the girl . . . she must be eight now. Is she . . .?'

'Normal?' Jenny said. 'Actually not. She's a gifted child, very gifted.'

'She gets that from Jed,' Barbara said, unwilling to grant Elaine anything.

'Yes,' Jenny said, 'I expect she does.'

'Was I ever really friends with her?' she asked herself as she went downstairs. However close they had been once, a universe lay between them now. But the Barbara she had just left was somehow different from the self-satisfied Barbara of Jersey days. 'I don't believe she is truly happy,' Jenny thought. 'But, God forgive me, I don't really care.'

'All right?' Dave asked, as they came in sight of the Cleveland Hills.

'Yes, I'm OK,' Kath said and smiled to prove it. Behind, in the back seat, L.J. was reading a story, putting her heart and soul into the different characters, holding the little ones on a thread.

'She'll be an actress, that one,' Dave said, trying to keep the pride out of his voice.

'We'll see,' Kath said. 'As long as she's happy.' She looked at the hills. Once beyond them they would be in alien territory.

'We're going a long way, aren't we?' Dave said and there was a tremor in his voice she had not heard often.

'Yes, we are.'

'I didn't push you into it, Kath, did I?'

'It'd take more than you to push me, Davy Gates. We're going to make a good life for our kids . . . and ourselves. I'm still young enough to be selfish. And if it doesn't work out, we'll come back. The north-east'll still be here, and we'll still have a place in it if we want.'

She felt like crying but she knew Dave needed her not to. Not now. She would cry later when there was no one to see. But she had no regrets. She remembered that day at the bus stop, Dave jogging by and Jed teasing her. Was it then she had fallen in love? Or the day they'd first made love? It was unimportant, for she'd certainly loved him ever since.

35

15 March, 1989

Steven allowed himself to be persuaded to watch the filming. 'Oh, come on, Mr Lockyer,' the young girl muffled up in the padded jacket had pleaded.

He had managed to hide his almost overwhelming curiosity, so he was able to heave a sigh and say: 'All right then, if I can't get out of it . . .' Cissie was busy running around like a mad thing laying out food and couldn't be prised away, so he went alone and stood behind the camera, watching.

It surprised him that there was only one camera, one man with a microphone that looked like a big sausage, another one setting up lights in the house in broad daylight, and one who seemed to do everything and nothing. And when Keir started spouting Steven felt a stir of pride that he had not felt for a long time. His son could talk! The words just flowed out of him and they sounded all right.

He called the miners the salt of the earth, talked about graft and sweat and the nation's debt to men who toiled underground. But did he mean what he said? The second and third time he did it the words were exactly the same, which made Steven suspicious.

Even the smiles came in the same places, and the little turn of his head back towards the pit.

They stopped then and did something called reverses, the girl taking off the padded jacket to go in front of the camera, repeating her questions but getting no answers, spending a long time nodding and half-smiling, like a toy dog in a car's rear window.

Steven had expected them to explore Belgate, but when Keir's bit was done they went into the pub.

'Vox pop,' said the girl, and then: 'We just want to ask

people what Keir means to them. And then we want to talk to you and his mum back in the house, after that lovely lunch she's dreamed up.'

They filmed Keir walking up the back street and into the yard three times, stopping people at the top and bottom of the street and telling them not to look at the cameras. A crowd of giggling kids formed and began to cat-call, and the man who did the lights chased them off with a stream of threats that was bound to bring them back for more.

'I think we'll break for lunch,' the girl said then, and they all trooped into the house.

'My mam's bread,' said Keir, throwing back his head in the kitchen and sniffing the air. Steven had never seen him do that before but the bossy girl seemed pleased.

'We'll get that after lunch,' she said. 'I love it.'

They filmed in the kitchen all the afternoon, repeating actions over and over, moving furniture so that the room looked nothing like normal, zooming in on Cissie doing simple, everyday things as though she was performing brain surgery.

Then they were gone, promising to come back next day.

'How long will this programme be?' Steven asked. Keir was staying behind for an hour before joining the rest of them in their Sunderland hotel.

'Half an hour, dad. Minus commercial breaks.'

'Half an hour! They've been at it all day, and they're coming back!'

'It'll all be edited down, you'll see. And a lot of what I say will be voice-over. That is, they'll show the pictures they've taken of you, mam, me walking round Belgate . . . and you'll hear me talking over them.' He laughed at his father's outraged expression. 'I know it sounds crazy but it works. You'll see.'

He stood up and picked the photograph of L.J. from the mantelpiece.

'Is this Elaine's little girl? She's pretty, isn't she. How old will she be now?'

'She's nine in August,' Cissie said proudly. 'What a little character. And clever! They said at the school that she was . . . well, some way of saying very, *very* clever.'

'Well, she would be,' Keir said slowly. 'Probably takes after her father.'

'Did you know the father?' Cissie was surprised. 'No one round here knows who the father is, not even Kath Botcherby . . . and she's her mother now.'

'Oh, I didn't know him. Not exactly. He was supposed to be bright . . . someone told me that once.' He put the photograph back and looked at his watch.

'I thought you might come down the club,' Steven said in what he hoped was an off-hand tone.

'Not tonight, I'm afraid. We need to talk, you see, about tomorrow's shoot. I'll try and make it before I go back, though. How is the old place?'

'Canny. Needing funds, though.' Steven took a breath. 'As a matter of fact they were wondering if you might come up and do a night . . .'

'I'd like to, dad. I'd really like to come . . . do a question-and-answer session or something like that. But it's out of the question at the moment, I'm afraid. I know it's hard for you and mam to grasp, but I'm not a free agent any more. When I move about it costs money. One day perhaps . . . In the mean time . . .' He was fishing in his inside pocket. 'Here's something for the funds. You shove that in and give them my regards.'

The twenty-pound note was crisp and unused, and it burned Steven's unwilling palm like a brand.

'You'd think he'd've stayed with us,' Cissie said, when Keir was gone and they had made their way up to bed. Steven agreed but didn't say so.

'He's not a bairn any more, Cissie.'

'I aired his bed and everything.'

'I know.' He wanted to reach out and comfort her but it might lead to other things. He lay listening to the tick of the clock. He liked silence more and more nowadays. Life pressed on too hard sometimes. You needed time to think, to remember. He thought a lot about the pit now, the drip of water, the squeaking of mice, the thunder of work at the face, the scent of an orange coming sharp along a shaft. They had cut him off from his work too soon, made an old man of him. He could still have been hewing coal. Coal! It was funny stuff. Black diamonds. It had cascaded along the conveyor, shining like a pirate's treasure. And someone's blood on every sodding lump.

Cissie was turning, putting her hand on him, touching him gently, putting her lips to his neck.

'Behave now, Cissie.'

'What's the matter, Steven? Don't you love me any more?'

He felt tears prick his eyes. How could he explain that he wanted her but he was too old, too tired, too played out to do anything about it.

'So how have you been?' Bart said, moving the single rose in its crystal vase slightly to the left so that it didn't obstruct his view of Jenny. His call had come out of the blue:

'Meet me for lunch, Jenny, I want to talk to you. Please.'

Jenny was here now because it had seemed wrong to refuse. They had been lovers for a long time, and even though her disillusionment had been profound she could remember the time he had made her happy. So here she was, in an expensive restaurant, wondering what he wanted to talk about and how they would fill the time till she could make her farewells and walk out of his life for good.

'I've missed you, Jenny. We were crazy to split . . .well, *I* was crazy. My fault. Still, you didn't take long to get over me.'

She contented herself with a smile, wondering what was coming next.

'He's doing well, isn't he? Your Paul.'

'My Paul? What do you mean?'

'Paul Kovalski, the Labour Party's rising star. You and he were very close, weren't you?'

'I knew him, we made programmes together. I like him and I admire him very much.' Jenny was suddenly afraid, pressing her knees together under the table, feeling the pâté turn to sawdust on her tongue.

'More than admire, Jen. You were in love with him if my information's correct . . . and he was besotted with you.'

Jenny knew what was coming now and knowing made her braver. 'Get to the point, Bart.'

'No point, Jenny . . . I'm not after blood. I want some co-operation, that's all. I'm doing a profile. The drums say he's one to watch, so I'm watching. He's got a good chance of being in the Shadow Cabinet in October, and you're part of his story, whether or not you like it . . . whether or not I like it. I don't, to be honest. I could kill the swine for taking you away from me, but that's beside the point.'

'If that's going to be your level of accuracy, Bart, it won't be much of a profile. You and I had been finished for months when Paul and I . . .'

'So you admit you and he were an item?'

'I never said that. I know him . . . I knew him . . . professionally.'

'And in the biblical sense, Jenny. He spent most of the period 1985-9 living at your flat in Humber Mews. Now don't look outraged, I didn't ferret that out, it's already on the files of every newspaper. The man's a politician, Jenny. Did you really think you'd be entirely unobserved?'

'So we're on file,' she said, marvelling at how quiet her voice was when she wanted to scream. 'What are you going to do about it?'

'As little as possible,' Bart said. 'That's the purpose of this meeting.' He put out a hand to take hers but she snatched it away, her face stony. 'All right then, Jenny – his affair with you is important. If I choose I can blow it up into a half-page story. All the details are there, and you're a media person. Punters love that. I don't, for personal reasons, want to splash you across the papers but I'm a journalist first, last and always, you know that. If I'm to play down such a newsworthy part of my story I need something to fill the space. You can give me that, Jen. You lived with this man, you know him. You know his standards, his ambitions, his views of key party figures . . . you know what makes him tick. Give me all that, and I'll go easy on his relationship with you. I can't promise to drop it altogether, but there are ways. "His friendship with Jenny Sissons, a BBC producer, is based on their working association in programmes like . . ." You know the formula.'

In the cab on her way back to work Jenny moved to get out of the line of the rear-view mirror before she let the tears flow. She wasn't important enough to be in this situation, surely? This was what happened to other people, people who knew how to handle it. When the cab drew up in Portland Place she knew she couldn't face her office. 'Take me to Humber Mews,' she said. 'I've changed my mind.'

Inside the flat she walked the floor, wanting to talk to someone but not knowing who. If only Euan were here – or Kath. But they were too far away to help, and her parents were out of the question. They had known there was a man in her life, a man called Paul, and they had probably accepted that she slept with him. Her mother had often hinted at how nice it would be when they married. They could never have imagined that he was already a married man and a public figure. To find out now would break their hearts.

But they *would* find out. Bart's sleazy article . . . for it would certainly be sleazy . . . would drop on to a million or more mats one Sunday morning. It could only be a matter of time till someone flaunted it under her parents' noses.

'I'm a grown woman,' Jenny told herself. 'I have a life. I can't help it if they get hurt – it's the way the world is.' But she could help it, or at least she could cushion the blow. If she gave Bart what he wanted – the treacherous dope on her former lover – he in return would play down her role. He would keep his word, she knew that, if only because he might one day be able to come back for more. She would be sliced, but not to the bone, and then put back on the shelf for later.

At seven o'clock Jenny rang his number, half hoping he would be in a pub somewhere and not answer.

'Bart Conroy.'

'It's me, Jenny.'

'Jenny! I was thinking about you.'

'I'm making one last appeal to you, Bart. I can't believe we had that conversation today. I can't believe you'd do this to someone who . . .' She couldn't find the right word. Had she loved him once? It was hard to believe that now.

'Jenny, Jenny, you're not thinking straight. You know this business. If I chuck the idea, someone else will pick it up. That's why *I'm* doing it, so that I can protect you as much as possible.'

'Liar, liar!' Jenny thought bitterly as she put down the phone. She knew how Bart worked, snouting for his own topic and working it up until it was irresistible to an editor. He could drop this idea if he chose – but even as she acknowledged that he could, she accepted that he would not.

36

17 April 1989

'So that's how I see it, Keir. If you follow those lines I think we'll hit the right note.' The woman rose from the couch. 'I'll just go and see if the set's ready, and then we can go ahead.'

Keir had come to audition for a new programme to be produced by an independent company for Channel 4. It was small beer, but at least it meant he could say he was doing something. He was starting to sweat now when people asked what was in the pipeline. The television profile was going out tonight, but he couldn't live on that forever, and he'd waffled about the book for so long that it too was old news. Not that he wouldn't do it one day – he would. It was finding a publisher . . .

He looked up as the other half of the company came in, a tall, gaunt woman in dungarees, looking like Annie Love without the designer label.

'Keir . . . long time no see! Has Bev explained what we're looking for?' He assured her Bev had explained, but that didn't stop her giving her version, which was diametrically opposed to Bev's outline. God save him from bitches! One said go right, the other go left . . . well he would do what he had always done, go his own way. He knew more about presentation than the two of them put together. And if he lost the piddling little job, so what? The profile tonight would put him back in the swim. They hadn't let him see the finished product, but according to them it was ace. He would be able to pick and choose after that, and sod the lot of them! His face was still smarting from the session on the sun-bed, but Keir knew he looked good. His hair had cost a fortune this time, needing treatments to restore its tone, but the finished effect was worth it. He had seen it in the admiring glances since he left the salon.

He followed the dungarees into the studio, seeing the tacky set, the sparsity of equipment. So that was how it was. Holy shit, why did Maggie Thatcher think more television meant better? Even a gravy train could be thinned down to nothing.

He smiled at the mock-interviewee to put her at mock-ease. The girl was obviously a typist called in to act as a sounding-board.

'We'll go in three, Keir. Turning over. Three, two, one . . .'

Keir looked into camera and smiled for a count of two. Get them gripped, sock it to them, don't forget the wink at the end. That was his formula. It had never failed him.

'Hello. I'm Keir Lockyer . . . and today I'm talking to a fascinating person.' If only the interviewees they gave him had been up to the standard of the interviewer the Keir Lockyer Show might still be the hottest thing on TV.

Jenny looked at the progress chart on her office wall. Five more programmes, then she was on leave for three weeks, and then off to the States, where she had an attachment to a radio station in New York while her American counterpart would work in BH. 'A cultural exchange,' it was called. Jenny doubted that there would be much exchange of culture, but she could learn a lot from other people's broadcasting techniques.

It would be Christmas when she got back to Britain. If she hadn't had this ache in the pit of her stomach over Bart, she would have been looking forward to it, making plans. She couldn't stall Bart any longer and she still didn't know what to do. She had alerted Paul and seen his terror, but he had no answers except to say, 'I trust you, Jen.'

She tried to concentrate on forward planning: Risley was all tied up for this week, Chinese dissidents for next. Trying to run the programme now, with the whole world in a state of flux, was not easy. Normally Jenny enjoyed a challenge but at the moment all she could think of was that she and Paul were going to pay a terrible price for the happiness they'd had together. It seemed that he was never out of the papers now, never off the box, his voice expressing an opinion each time she switched on the radio. He was hot news, and if Jenny hadn't been involved she'd have admired Bart for homing in on him. How different it was to be on the receiving end of media interest. Had she ever crucified anyone as she was being crucified now?

The phone ran and she lifted it, both hoping and fearing that it would be Bart.

'We've someone for you down in Reception, Miss Sissons. A Mr Lockyer.'

'Keir!' Jenny said when she reached the foyer, holding out her hand and accepting his kisses on each cheek.

'I was in your neighbourhood – I've just come from a pre-record, well, a pilot, actually – and I suddenly thought, "I want to have lunch with Jenny!" Are you free?'

Keir had kept his cab and they sped down side streets to a quiet shadowy restaurant he knew. If Jenny had been able to think of an excuse she'd have said no, but when they settled at a table she was glad she'd come. Keir was at his most entertaining, on a high about a profile of him that was going out tonight.

'You mustn't miss it, Jen. Heaps of shots of home. You'll be nostalgic. Well, you know me . . . I'm a Durham boy, always will be.'

They were on to dessert when Keir mentioned L.J. 'They've got a huge photo of her above the fireplace. They worship her. Mam says she's bright?'

'More than bright. Her IQ is huge, apparently.'

'And Kath and Dave have taken her to wallaby land?'

'She's part of their family now, Keir.'

'I know, I'm not quibbling. Kath's been a brick. I just thought . . .' He was toying with his fork. 'Was Jed L.J.'s father?'

'Why do you ask?' Jenny was suddenly frightened. What did Keir know?

'No reason. Mam said it's her ninth birthday in August . . . I suppose I did some sums. No reason, just sheer idle curiosity.'

He was suspicious – he'd been told L.J. was something special and all of a sudden he'd done some sums. She must keep calm, but *did* she have a right to lie? Did L.J. need Keir?

'After all, Elaine and Jed were thick but not *that* thick that early on,' Keir continued. 'And no one has said anything to Jed's people, which is odd.'

'That was what Elaine wanted, and she was the only one with any right to make decisions, Keir.'

His eyes had flickered at the mention of Elaine but that was all. He didn't give a damn! If L.J. had been an ordinary child he wouldn't even have noted her existence. What had Elaine said, so long ago, when they had talked of Keir? *'He'd make a lousy father.'* She had been absolutely right.

'Of course it was Jed's baby,' Jenny said firmly. 'He knew all about it, and they were going to get married. But when he died Elaine felt she couldn't burden his parents: they'd have wanted to help and they couldn't afford it. So she decided to stand on her own two feet. She was very brave.'

'I see,' Keir said, forking cheesecake to his mouth. 'Oh well, I'm glad to hear the facts.'

Steven heard Cissie at the door, breathless with haste.

'Have you not got the telly on yet? We don't want to miss it!'

'There's a good half-hour yet, Cissie. Three-quarters, more like. Get the kettle on.'

Cissie waffled on while she brewed the tea, and Steven knew something was coming. He was getting old, but he could still clock the workings of her mind.

'You know that dog down the street . . . the golden retriever?'

'The cross-breed golden retriever,' he countered.

'Well, call it what you like . . . she says they can't go on keeping it.'

'They haven't had it five minutes.'

'I know. They didn't bank on how much it'd eat. She says it knocks back a tin of meat a day and they can't afford it.'

'No wonder. And it's not full grown yet.'

'We could afford it, Steve.'

He mocked her cruelly, now, anger rising in him that she

couldn't see sense when it was staring her in the face. ' "We could afford it, Steve. Three bags full, Steve." No dog's coming in this house, Cissie. Not to be put on to me, and I don't see you taking it round the block at midnight. Now shut up, pour the tea, and let's get settled for this bloody film they made so much fuss about.'

Keir took the phone off the hook and carried his drink to the zebra-striped sofa. He would replace the receiver after the programme. People would be ringing then, both friends and enemies fulsome in their praise.

While he waited for what seemed an interminable time he thought of those watching . . . old friends like Jenny, former colleagues like Desmond, who hadn't lifted a finger to help when ratings fell, former lovers like Toni and Annie Love. He closed his eyes against an image of Annie naked in her bed. Instead he remembered the sweetness of Jenny. He'd been slow there . . . or too quick.

And Elaine – he had almost forgotten Elaine. Was it once, or twice? It had been the thrill of beating Jed to it, once he had seen how the land lay there. And yet Jed had left his mark. A child. It was a funny old world. The adverts ended and Keir put down his glass and leaned forward, not wanting to miss a single frame.

'This is Durham, the heart of the coalfield. The last place you'd expect to have spawned a media personality. Like him or loathe him, you can't have missed him . . . and this is where he was born, in this two-up, two-down colliery house.'

Keir felt a frisson of unease run up and down between his shoulder blades.

'His name is Keir Lockyer and this . . .' the girl leaned towards the camera and winked hugely . . . 'is his trademark.'

When had they filmed that? Not when he was there. They had never mentioned an introduction. Keir put a hand to his mouth, suddenly knowing he was going to be well and truly shafted, knowing it because he had helped to perfect the process.

*

379

Jenny opened Euan's letter as the titles rolled up. She didn't hear the opening remarks, so intent was she on what Euan had to say.

I know something's wrong, Jenny, I sensed it in your last letter and I want to know what it is. I'll ring you on the 17th, by which time I hope you'll have received this. Don't put me off, Jenny. If I don't get satisfaction I'll come over.

It would be a relief to confide in someone. There was nothing Euan could do, but it would help to tell him about Bart and Paul. Jenny put down the letter and turned her attention to the screen.

'He was like a breath of fresh air at first. A bit gauche . . . a little boy with a new toy. He could be outrageous but we liked it.'

Jenny recognized the TV critic of one of the tabloids, pink and perfect as she drove the knife home. She watched as Keir walked the Belgate streets, his voice talking in counterpoint to whatever he was doing. Now he was in a restaurant, Le Gavroche probably, seated amid the gleam of silver and crystal, a glass raised in his hand, while the voice-over turned the whole thing into a farce.

'I'm a socialist. What else could I be with a background like mine? I was named after a socialist hero, Keir Hardie, remember. I'm happiest in the coalfield, sharing with the people of Durham, my people . . .' A waiter moved into shot and began to serve escalopes from a tray. This was not a profile, it was an assassination. Jenny reached for the remote control but she couldn't switch off, couldn't miss one gripping second of the kill.

37

1 May 1989

Cissie's tongue had lashed him out of bed when he had wanted nothing more than a bit of peace. 'You're losing sight of yourself, Steven Lockyer. I don't think you even get washed now. A good Bank Holiday, and all you can think to do is lie abed.'

He had clashed from the house then, buttoning his clothes as he went, shivering a little as the wind got to him. There was still a sting in the tail of winter, in spite of the blossom on the trees.

It was quarter-past eight, nowhere open for a man to go to. He passed by the Aged Miners' cottages, seeing the windows still eyeless and curtained, the doors bolted, the milk on the steps. He would end up there, he and Cissie, gazing into their banked-up fires till they went glakey and were carted away.

Cissie had called after him, scared of the effect of her words. Well, let her worry! It would do her good to have a fright. But that meant he would have to stay out for a few hours, and there was only the beach. He would find a place out of the wind and pass the time as best he could.

May Day, the Workers' Day, and he had been driven from his own home – that was what he had come down to. Steven turned up the collar of his donkey-jacket and made for the promenade.

He was half-way down the steps when he heard the screaming and saw movement above the water's edge. Not a body, surely? Whatever it was was moving, and the yapping and screaming was getting louder.

Steven crunched across the pebbles at the tide-line and then crossed the wet sand, feeling his footprints fill instantly with water that seeped into his shoes with a squelching sound. The sand dragged at his feet until the muscles of his calves ached with the effort of moving. He hadn't hurried for a long time.

It was a seal, a pup still to judge by the size and the coat. Its back flippers were bound tight by yards of netting and it was trying to get up the beach to some kind of safety. The huge black orbs of its eyes swivelled desperately, begging him to help. He pulled at the net, knowing as he did so that nothing would part that wicked tangle except a knife.

'Lie still, bonny lad. Lie still. I'll be back in a minute.'

He turned and started back up the beach, the wet sand sucking at his feet as he ran. If lads came down . . . well, you never knew with young'uns. And any minute now the dog-walking brigade would be out in force. He thought of the seal's imploring eyes and quickened his pace.

There were only the aged miners within reach. If he passed them and went on into the village it would take too long. He ran over likely ones in his mind. Jack might have a Stanley knife, and others would come and hold it still while he cut . . . surely everyone would help?

He knocked on Jack Longman's door, his heart sinking as he heard the shuffling steps on the other side.

'There's a seal, Jack, a young pup . . . stranded on the beach, tangled in nets. Down from the Farnes, likely. I'll need a hand to get it back in the water . . . can you come? And bring a Stanley knife, if you've got one?'

'Aye, I'll come, lad.' The gnarled fingers were already fumbling at the shirt buttons. 'But I haven't got a knife.'

Steven ran on to the next house but the gasping breath of the owner told him there would be no help there.

'Can you knock up next door and pass the word? Anyone who can help . . . I'm going back. We need a Stanley knife!' He had his old baccy knife in his pocket but that would never do.

His breath was coming harsh from his lungs as he reached the beach again. The tide was further out now, the seal still crying above the water's edge. He turned and saw a white-haired old man appear at the top of the cliffs and begin his descent. And then another, shrugging into a parka, and another, pulling a woollen hat down over his ears, something gleaming in his hand that might be steel. But they were old men, old like him – and a German Shepherd dog had appeared at the end of the beach.

Steven began to run, plunging through the dry sand, sinking in the wet, unfolding the clasp knife as he ran. The seal had seen or heard the dog and was turning back towards the water. If it took to the sea still tied it would die . . . from starvation or because a fisherman blew its head off with a shotgun. That was what they did with seals that menaced their nets.

An iron band had fastened about his chest but Steven didn't care. Only the young life mattered now. The seal must live! He saw the black eyes imploring him, the whiskered mouth open in a scream as the dog's massive paws thundered on the wet sand. He reached down and scooped the seal into his arms, holding it high above the dog's baying mouth, feeling old muscles stir and tear, and then hold steady.

The dog's owner was shouting but the animal continued to leap. It would have him over in a minute. He had dropped his knife when he picked up the seal or he would have driven it into the dog's huge head.

'Hold on there, Steve, I'm coming!' Jack Longman's wispy hair had floated from his scalp, leaving his head pink and vulnerable, but he was seizing the dog's collar. And then another old man was there, ferrety-faced with a cyst on his chin the size of a pea, but laying about him like a good'un as other dogs appeared from nowhere to join the pack.

Steven held the struggling seal high until his army of old men vanquished, and then he laid the pup down on the sand and they held it while he sawed the net free with a Stanley blade. It looked at him once, whimpering a little until it felt its back flippers free. Outside the ring of old men the dogs still menaced. If he let the seal go free they would tear it to pieces.

'Come on, lads! One good heave.' But they had shot their bolt, the lot of them. It would be up to him. Their arms were willing enough, but his were the only arms with the strength – and he wasn't sure they had strength enough.

He kneeled down, as he had kneeled many a time in the pit, and put his hands to the smooth flanks.

'One, two – three!' He felt the iron band tighten across his chest. Muscle strain or heart attack? Never mind, if the seal went free.

He was up and someone was whooping like a little bairn. He felt his heart thundering wildly, his legs staggering under the weight, and then they steadied and he was moving, the dogs falling arse over tip towards the sea, the pup carried above them until he felt water up to his thighs and could let it go, suddenly galvanized, dipping and gliding away from danger to its natural element.

He felt the water soak up to his waist in that moment. Cissie would kill him when he got home.

'Your lass'll go leet about this,' someone confirmed and they all laughed but Steven stood in the water, feeling his strength return. My God, he was still alive after all, and at this rate he might live forever.

On the early morning TV news, Jenny watched the scenes in Prague as protesters gathered in Wenceslas Square. There would be trouble before the day was out – and if not here, then elsewhere in Europe. The Czechs were shouting for their hero, Gorbachev, and for the release of Vaclav Havel, the jailed playwright who had come to symbolize all their hopes for the future. She wondered if Euan was there, even seeking his face as the cameras panned over the crowds. He had tried to comfort her the other night on the telephone, but it had been to no avail. It was the third time he had rung since the night of Keir's disaster. She would have gone home this weekend, but she knew she couldn't hide her distress from her parents. Safest to stay away.

Bart had given her until Wednesday. Then he would go ahead, with or without her co-operation. 'I've got no choice, Jenny. A story's a story. You know the score.' And the worst thing of all was the knowledge that once she had loved him, come alive at the touch of the man who was going to betray her. It would ruin Paul's marriage, and probably his career. And break her mother's heart. They were proud of her; they trusted her – and they could no more disregard her featuring in a tabloid sensation than fly.

Jenny's bell rang and she pulled the belt of her robe tighter, wondering who could be calling on her so early on a Bank Holiday.

'Can I come in?' Alan stood on the step, dressed in beige slacks and a blue sweater.

'Alan! How nice.' She stepped back, wondering how she would get through the next few hours.

'It's OK, Jen, you don't need to put a good face on. Euan's told me the whole story.'

Jenny turned away from him, wishing with all her heart that she had kept it to herself. She was going to cry, and if she cried Alan would despise her. Worse, she would despise herself.

His hands were on her shoulders, turning her round to face him, letting her put her embarrassed face against his sweater. 'It's over, Jenny. No more worries. So have a good blub and then forget it.'

'What do you mean?' She leaned back to look up at him incredulously.

'I mean it's over. I've seen your ex-friend . . . and I do hope it's ex . . . and he won't be proceeding. Not as far as you're concerned, anyway. Mr Kovalski must take care of himself, but *you* won't figure in any article.'

'What did you do?' Jenny's voice was anguished.

'I didn't do him grievous bodily harm! Give me credit for some subtlety.'

'Tell me what *happened*!'

'I will. But fair dos, I rushed round to tell you before I had breakfast. Get the kettle on, poach me an egg, and I'll cough the lot, guv. I rang last night, but you were out.'

Alan sat at her kitchen table while Jenny moved between fridge and stove, and told her of his meeting with Bart the night before. 'I arranged to see him in my chambers – that was one stroke to me from the start. I told him that if he included you in his sordid little piece you would sue him for libel.'

'But how could I?'

'Oh, I freely acknowledged you hadn't a leg to stand on.

But I would appear for you for free, I told him, while he'd have to engage counsel at enormous cost, with no newspaper to foot the bill. I told him we'd lose, but that before the verdict I would have pilloried him as a man who ratted on a former lover. The British public don't like cads.'

'Thank you,' Jenny said faintly. 'I can just hear myself being questioned about *two* liaisons!'

'Oh, it wouldn't have come to that,' Alan said airily. 'But wait a moment. I said we'd lose but he'd be very lucky to be awarded costs, and cited cases to prove my point. And I suggested that if he waited he'd probably catch Kovalski *in flagrante delicto* with someone else . . .'

'That's not fair!'

'That's *very* fair, Jenny. My brief was to get you off the hook. MPs are fair game.'

'So Bart agreed?' She couldn't argue with Alan, not when he had done her such a favour.

'After a while. I simply helped him to see that the game wouldn't be worth the candle because I'd make it impossible. And there was a final clincher . . .'

'What?'

'I told him that if he went ahead I'd beat the living daylights out of him. Now, is that egg ready?'

Jenny cried a little as she cooked Alan's eggs, tears of relief and humiliation and gratitude.

'That's right,' Alan said comfortingly, buttering his toast. 'Get it over and then put it behind you.'

He drank two glasses of orange juice and ate eggs and toast with relish, persuading Jenny to eat a slice with butter and marmalade. Then they sat on over coffee, talking easily of past and future.

'I might go into politics, Jenny. It's no good just talking about an unfair world, you have to make things happen. I thought David Owen might do it in the middle-'80s but that was just a dream. It depends what happens in Britain in the next twelve months. When you see the furore in Eastern Europe at the moment, you realize nothing is forever.'

Jenny looked at his resolute face. 'You never know, Alan, you might just do it one day. Will I be welcome at No. 10?'

He smiled. 'That's not for me, Jenny, I don't care for

limelight much. I'm a foot soldier, but they're a vital part of any army. Now, it's eleven o'clock on a Bank Holiday – if you're free, I'm free. What say we go to the funfair on Hampstead Heath and eat hot dogs and forget what worthy citizens we've become in the last ten years?'

While Jenny dressed she found herself thinking suddenly of what Kath had said about Alan being good in bed. She put her hands to her flaming cheeks and wondered how the day would go. One thing was certain; tonight she would go down on her knees and give thanks for having him for a friend.

Bart's article on Paul would not go ahead now. There had been no other woman, whatever Alan might think, and without a sexual element there would be no real mileage in Paul's story – or not Bart's sort of mileage, anyway. She clipped in looped ear-rings, thinking of Bart having the frighteners put on him. Poetic justice. If only she could have seen Alan in action!

It was then that she thought of Jed's death. It had been Alan on the bank, she was certain now. Not walking away callously, but doing it because there was no hope and it made no sense to lose two lives instead of one. She picked up her shoulder bag and went back to the sitting-room, the euphoria she had felt earlier a little diminished by the realization of just how single-minded Alan could be.

38

21 June 1989

Jenny caught an early train from King's Cross, settling into a window seat, gratefully drinking terrible coffee from a paper cup. It was Midsummer's Day and she was going home. At eight o'clock she would take up her vigil on the hill – who else would be there to join her? Not Kath, but she *had* remembered. Her letter, and photographs of her laughing children, were there in Jenny's bag. *'Give my love to the hill and all its shades, Jenny. A part of me will be there always. I will not forget.'*

No word had come from Euan, and that surprised Jenny. He was ecstatic now about the changes he was seeing in Berlin: perhaps he couldn't bear to leave? She couldn't believe Keir would be there, either. He seemed to have faded from sight. And hopefully Barbara had not remembered: 'I couldn't cope with Barbara,' Jenny thought. Alan would probably have come, but he was away at the Dorminster Assizes.

At least she could say her goodbye to the town she would not see again until she came back from the States. She finished her coffee and shook out her paper, seeking news of China. The government was still clamping down on the heroes of Tiananmen Square, showing them bound and cowed before their execution. Yesterday she had walked through London's Chinatown, where the casualty figures were pinned up on the wall, seeing some people weeping and all the rest with misery in their eyes. She had lunched in a restaurant that normally buzzed with chatter and laughter, but the diners had been strangely subdued, as though the dark arm of the Chinese hierarchy over-shadowed them even there.

At Peterborough, Jenny went in search of more coffee.

There was an elderly woman at the buffet whose face was slightly familiar, but Jenny couldn't place her. When she had been served she went back to her seat, looking out at the countryside as she drank, and thinking how beautiful Britain was. Why was she leaving it? Still, six months was not a life-sentence and she would learn a lot.

She had decided to stay in radio when she came back. She had been offered an attachment in television but it was not for her – look what it had done to Keir. Apparently he was a wreck now, begging anyone and everyone for a job. Did TV change him or had he always been a cardboard figure? Could you be plucked from obscurity and be lionized, and come through it unscathed? There was so much about television that was good; it had such power to inform and entertain. But wherever there was a power for good there was a power to destroy, and Keir had fallen victim.

Jenny tried to feel sorrow for him but it wasn't easy. She really didn't care about him one way or another. He had taken her virginity, and now she could hardly remember his face. How many women would tell the same story, if they were honest?

She could remember Alan, though. Since the day he had come to tell her she was free of Bart they had been out together three times. Pleasant, platonic outings. Well, almost platonic. He had kissed her goodnight on each occasion – but he had been doing that for years. 'We're friends,' Jenny told herself firmly. 'Good friends.' But whether or not his kisses were platonic their effect on her had been more than a little disturbing. Had there been more than friendship in his embrace? 'I want there to be,' she acknowledged at last and smiled suddenly, thinking of Kath. If she were here now her eyes would have popped. 'You fancy him,' she would have accused, and gone on affirming it in spite of Jenny's denials. 'But does he fancy me?' Jenny asked herself. There was no answer. With Alan, it was almost impossible to know.

He would have come to the hill if he could, she was sure. She drained her cup and put it down, reaching in her bag for her paperback. A good read was just the thing to dispel disturbing speculation.

'I thought it was you.' The elderly woman from the

389

buffet was looming over her. 'I saw you up there and I said to my friend when I got back to my seat, that's the girl I saw at poor May Denton's funeral.'

'Of course,' Jenny said. 'I remember now.'

'Have you seen Alan lately?'

'Yes, I saw him last week. He's at Dorminster Assizes this week but he still goes home regularly. He means to keep the house.'

The woman was nodding. 'Poor May. She loved that house.' Why did she keep saying 'poor May'?

'Did she have a sad life?' Jenny ventured and the woman dropped into a seat and launched into her tale.

'May was in service when she was a girl. We came from a nice family, but there was no money. I was her cousin, you know. She went into service down in London, but I found a place in Darlington so I could get home a bit more often. Poor May, she hankered for her home. When the war came she went into munitions, and afterwards, when her dad died, she wanted to be back beside her mother so she got a place in Sunderland as housekeeper to Edward Denton. Oh, he was a gentleman! Never been married . . . worked in a solicitor's office. Of course he was retired by that time. May lived out and looked after him in the day. He was sixty-five, she was twenty-eight. I suppose one thing led to another, and he married her. We warned her May doesn't mate with December, but she wouldn't listen. It caused terrible talk, but they knew it would. They were that wrapped up in each other I don't think they cared what other people thought.

'And then she fell with Alan. They were that pleased, like a couple of children . . . until the letters started.'

'Letters?' Jenny said. 'You mean anonymous letters?'

'Filthy letters,' the woman said. 'Too much for Edward. They made him feel dirty, you see. Just for loving. He loved the boy but it broke his heart. He died, and May was left. She didn't trust anyone any more, cut herself right off. All she cared for then was the boy . . . pushing him to make something of himself.'

'Why was that so important to her?' Jenny asked.

'Because of what the letters said. That the child would be an imbecile . . . a monster because of having an old man for a father. She wanted the world to see what a fine boy he was, to show them they were wrong.'

'He's a fine man,' Jenny said. 'I'm proud to have him as a friend.'

'Well, I don't know if Alan knows or not. About the letters, the trouble. If he doesn't I wouldn't let on, although it was common knowledge at the time. And now I must get back to my friend. Tell Alan I was asking after him. I was his mother's cousin . . . I don't know what that makes me to him.'

How strange Alan's life must have been, Jenny thought as the train pulled into York. And how lucky she had been to be born into a happy home.

Steven watched television with one eye while he tried to fasten his boots. The hippies were at it again at Stonehenge, flocking to celebrate the summer solstice. 'What d'you think of them, Cissie?' he asked as she came into the kitchen.

'Crazy,' she said amicably, filling the dog's bowl at the tap.

'That's right, get a drink,' Steven said, moving to look down at the golden head as it lapped thirstily. 'And then we'll go on a good hike. Get a bit fresh air.'

He stood beside Cissie, breathing in the soapy scent of her, then he put up a hand to cup her breast.

'Now don't start,' she said but she moved against him just the same, both of them remembering how sweet last night had been.

Out in the street, the dog scurrying ahead to sniff at corners, Steven thought about the approaching summer. Cissie was hoping Keir would come home: she always would – mothers were like that. But he had come to terms with the fact that his son had grown away from him, and he felt better for the acknowledgement.

In sight of the sea, he scanned the beach, wondering, as he always did, if the seal was out there, dipping and diving, glorying in its freedom.

Cissie had a leaflet about a club you could join for cheap trips to Australia. It would be a big thing, but Kath was always saying, 'Come' – even little L.J. asked 'When?' in her letters. And it would give him something to talk about when he got back, by God it would!

'We'll see,' he told the dog, as he perched on a rock, and fondled its great head. 'We'll see.' And then he stood up and began to walk, lifting his face to the sun as he went.

'Have you chosen?' Annie Love asked, looking down at her own menu. She wore huge glasses now and there were new, deep lines on her face. Keir scanned the menu, hardly able to believe that he was here, in such an in-place, by her invitation.

'The crevettes, I think. Please. And the roast duck.'

'Right,' she said, when the wine had been tasted and accepted, 'let's get down to things, and then we can relax and enjoy the meal.'

Keir nodded, wondering how he could have so misjudged Annie. She was going to forgive him, take him back and make him a name again. She had gone to a satellite TV company but she hadn't forgotten him. She was a truly magnanimous woman.

'I want you for this show because it's right for you. In fact I dreamed it up with you in mind.'

'Has it got a title yet?'

'"Pay the Price" – that's what we're calling it. "Pay the Price". It will be glitzy and over the top . . . if we're going to win a huge audience, it has to have an extra dimension. As for money . . .'

Keir tried not to show he was holding his breath. This was where Annie could settle old scores. He was on his uppers, and she knew it. He had to take the job, even if she screwed him over the fee.

'The money will be good. More than the old days. So good, Keir, that you won't be able to turn it down.'

Jenny left the house at ten to eight, the dog running ahead of her, still lively now although it was getting old.

She climbed to the summit of the hill and stood looking down on the town, thinking how it had changed since that morning when they had watched it wake to a new decade. The shipyards were silent now, and unemployment had reaped a sad harvest, but the heart of the town still beat strongly. It would survive.

It had been a difficult ten years, for the town and for the world. 'It's going to be a decade of health,' Jed had said and yet Aids had stalked across five continents. They had been wrong about many other things, too. Jenny tried to remember the statistics Keir had rattled off. '*One of us will be dead*,' he had said, but two of them had died. '*And three will be married*,' but it was only two. They had believed they would all be around to climb the hill at the end of ten years: in fact there was only her. Kath and Euan were in far-flung countries . . . and Keir and Barbara were perhaps furthest away of all.

Jenny lowered herself to the grass, thinking as she did so that in a way Jed was more alive than any of them. He was here on the hill in the waving grass, forever young, never knowing the defeat of living. What had the vicar said at his funeral? '*He understood the gift of life . . .*'

'Jenny?' Euan had come up behind her, walking soundlessly in the soft, summer grass. Even the dog, nestled at her side, had been oblivious of his coming.

'Euan!'

He dropped down beside her and Jenny kissed him warmly. 'I wondered why I hadn't heard from you. It's so *good* to see you, but I never expected it. What a wonderful surprise! Where are you staying? . . . have you been home? . . . have you eaten? . . . did you come by car?'

He stayed her questions with a smile and a shake of the head.

'I'm OK. I had to come, because I knew you might be here alone.'

Jenny told him of Kath's letter and Alan's Assizes. 'As for Keir and Barbara, I don't suppose it's even occurred to them. I'm off to the States at the weekend, as you know, but I'm glad I could be here. And *so* glad you've come too.'

The light was softening now and the birds were ceasing to sing. In an hour or so dusk would gentle the landscape and then the lights would spring up.

'Do you remember the lights that night?'

Euan nodded. 'I remember it all, Jenny. I'll never forget it . . . or what followed.'

He meant Jed. She knew it and reached out to touch his hand. But he withdrew from her, only his eyes meeting hers, holding her gaze so that she could not look away.

'It was me with Jed, that night, Jenny. Me on the bank when he died.'

Jenny heard herself laugh nervously. 'Don't make jokes, Euan. You were miles away.' Her mouth had gone suddenly dry and her heart was thumping unpleasantly in her chest. She couldn't believe what he was saying. She mustn't believe it because she couldn't bear it to be true.

But Euan was speaking again, forcing her to accept what he said. 'We'd made an arrangement to meet to say goodbye. He knew I felt dreadful about going away. "We'll get rid of the others," he said, "and have a good natter before you go." You know what he was like, always giving to the one who needed him most. I made my goodbyes and drove off, but then I doubled back and waited for him down by the river. He walked uphill and away from you, so no one would know. He wanted to cheer me up, Jenny, so when we got there he started larking about. He was out on that blasted tree . . .' Euan's voice broke suddenly, but before Jenny could speak he was going on.

'He walked along it once . . . almost ran . . . and came back. He was full of good spirits. He went out again, further out this time, and then he turned. "I'm going to get married," he shouted. "You'll have to come back from Rhodesia to stand up for me, you old sod. I'm going to marry Elaine!" I could hear what he said but it was as though it was from a long way off. I felt such *jealousy* I thought it would choke me!'

'Oh, Euan! I knew you cared for Elaine, but I never realized how much.'

His smile was the sweetest she had ever seen. 'It wasn't Elaine I loved, Jenny. It was Jed. I loved him from the day I met him, and I always will.'

Jenny couldn't take it in for a moment. But Euan was talking again and she tried to concentrate on what he was saying.

'I turned and walked away and that's when I think he must have felt the tree move. He started to shout: "Help! Help me, Euan." But I thought he was larking on, the way he always did. I walked away. I walked away and left him. All I could think of was getting back to the car to lick my wounds. The only person I will ever love, and when he needed me I was walking away.'

They stayed on the hill while dusk fell and stars started to come out, not talking much because there was so little left to say. In their long silences Jenny probed her feelings about Euan. Did it make a difference that he was homosexual? No. The only wonder was that she had not realized how total his love for Jed had been. Now, all that mattered was to bring him comfort, to console her friend . . . her best friend, apart from Kath – for Alan had moved into a separate category, almost without her knowing.

'I understand how you feel, Euan. How you must have felt all along. I only wish you'd let me share it with you sooner.'

Euan reached out for her and drew her close, as the dog's ecstatic circling of the hill grew slower and it began to tire of freedom. The stars twinkled above and Jenny thought once more of Auden's poem: *The stars burn on overhead, unconscious of final ends.*' There had been a final end for Jed and Elaine but for the rest of them there was the pain of life.

'If you'd stayed on the bank and tried to save him it wouldn't have helped,' Jenny said once, and Euan nodded.

'At least there's L.J.,' he said at last. 'At least we have Jed's child.'

Jenny let him go on taking comfort from the lie. Why take it away from him now?

At last the dog nuzzled her hand to show it was time to go home, and they began to pick their way down to the road.

'Take care,' she said at the foot of the hill, folding him in her arms to hug him better. 'I need you, Euan. We all need you. So be careful.'

His lips on her hair were gentle and when she pressed her cheek to his she felt his tears.

'Enjoy America,' he said. 'Don't forget to come back.'

Behind them the hill loomed, the most solid thing in her life.

'I'll come back, Euan. And so must you.'

EPILOGUE

Christmas Day
1989

Jenny stood on the hill, come back from New York to a northern town still somnolent from Christmas feasting. Here and there a child rode a new bicycle; a boy and girl, arms entwined, walked to visit parents, bearing gifts. It had been a good Christmas, a good end to a difficult decade.

In Europe barriers were coming down. East and West Germans were free to embrace, Ceaucescu was toppled in Romania, Vaclav Havel was soon to be elected president of the country which had once imprisoned him. Freedom was breaking out all over Europe, and please God, it would soon extend to other continents. 'They are calling this the "Quiet Revolution",' Euan had written from Prague. 'They aren't thirsting for revenge, just for a new and happier life.' It was impossible to be sad in the midst of so much rejoicing.

Jenny had arrived in Sunderland to find flowers from Euan and Christmas cards from Alan and from Kath, who had enclosed more snapshots of suntanned children and news of a fifth on the way. There had been a card from Barbara too, over-printed in gold and big enough to serve as a tray. Jenny would have laughed, but she had looked up then and seen Keir on TV – smiling as the contestant won and he, the presenter, 'Paid the Price', carrying out a forfeit guaranteed to make him look ridiculous. Smiling all the while.

Now, holding her thick coat tightly round her, she looked down on Sunderland again, trying to dispel that image. 'A nice district,' that's what her mother called it. 'A nice district' in a nice town. Her town. She was home and she should be happy. She *would* be happy once she settled

397

down and got back to work. She knew she loved Alan, but she had accepted that it would not come to anything, and she could even appreciate the pain of longing.

She looked around her. 'I would fight for this hill,' she thought remembering pictures of people battling for their country's freedom. One day she might have to fight, for all over Britain the green and quiet places were under threat, hedgerows and grasslands were gone or going, tracts of heath and moorland put to the plough so that flora died and fauna fled, perhaps never to return. And not only Britain was in peril. That would be the battle of the '90s, to save the planet. If it was lost, the people would perish.

She bent and touched the short winter grass. 'Merry Christmas,' she said softly – she didn't know to whom.

She whistled up the dog, then, and turned for home, stepping from tussock to tussock on the steep descent until she saw a man climbing towards her, his dark head bent. Her breath stilled in her throat until she saw that it was really Alan.

She closed her eyes then, squeezing the lids together in pleasure at the sight of him, until at last he looked up and saw her.

'I've come to find you, Jenny,' he said and held out his hand.